CONCEPTS IN ENGINEERING DESIGN

CONCEPTS IN ENGINEERING DESIGN

SUMESH KRISHNAN - DR. MUKUL SHUKLA

Notion Press

Old No. 38, New No. 6
McNichols Road, Chetpet
Chennai - 600 031

First Published by Notion Press 2016
Copyright © Sumesh Krishnan & Dr. Mukul Shukla 2016
All Rights Reserved.

ISBN
Paperback 978-1-945497-61-2
Hardcase 979-8-89519-019-7

Table of Contents

Chapter 1

Chapter Objectives

When you complete your study of this chapter, you will be able to:

- Understand the nomenclature of design terms.
- Understand Design, Engineering Design and Engineering?
- Give an Introduction to Modern Engineering Design
- Understand the key steps in the design process
- Explain customer's role in the design process
- Understand elements of design
- Understand types of design
- Understand Various ways to visualize design
- Designers vs, Scientist Method
- Understand problem solving Methodology
- Design engineering solutions to societal needs

1.1 Fundamental Understanding of basic design terms

Methods, Process, System, Analysis, Synthesis, Design, Engineering Design and Engineering.

- Analysis: listing of all design requirements and the reduction of these to a complete set of logically related performance speci-fications. Analysis is also called reduction, and "reductionism" to scientists means the importance of analysis.

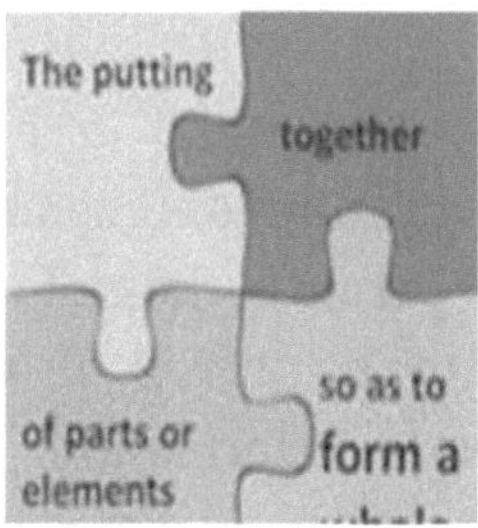

Fig.1.1 synthesis

- Synthesis: finding possible solutions for each individual performance specification and building up complete designs from these with least possible compromise.

- Evaluation: Alternate designs are evaluated to fulfill performance requirements of use and manufacturing of product final design is selected.

- System: Systems are universal in nature from home network to cyber system, from water purifier system to water treatment system, from railway networks to power grid system and from microbiological system to our solar system and so on. System study spans from engineering to biology to political and even financial system (this is an example of intangible system)

Commonly we say that "The system is fine," when things are done easily we say "the system is a failure," when it troubles us. The commonsense notion of system is also very familiar one "the system is a whole comprising interrelated parts with some complexity" but the system is not merely the sum of its parts. Commonly there is no restriction on whether the related parts, are people or things; are animate or inanimate, the parts are concrete or abstract; hardware or software etc. Nevertheless, their interrelations, although complex, should be open to some rational understanding

"System" is a buzzword in technical studies; the system may appear heterogeneous in differentiable stuffing only the boundaries separate the system from the environment. These days the most prominent notion of "system" is expressed by the metaphor of a "seamless web." Introduced by Thomas Hughes, the prime promoter of "systems thinking," it includes the notions of perfection, holism and resistance to analysis.

Systems study is very important in engineering. Engineers are responsible for designing and building communication, transportation, and other systems that operate in the real world. The systems concepts must be particularly to engineers of various field for example thermodynamic systems, information systems, power systems, communication systems and so on are very imminent in engineering studies.

- Inputs/Outputs may be
  - ◆ Matter (eg. Fuel or raw material)
  - ◆ Energy (Electricity or heat)
  - ◆ Information (eg.a bitstream or an operator keypress)
- The relationship between inputs and outputs are the system functions. According to Ludwig Von Bertalanffy (father of the systems approach) "In order to understand an organized whole, we must know both the parts as well as the relation between them."

What is a Process

- (Webster) A system of operations in producing something; a series of actions, changes, or functions that achieve an end or a result
- (IEEE) A sequence of steps performed for a given purpose

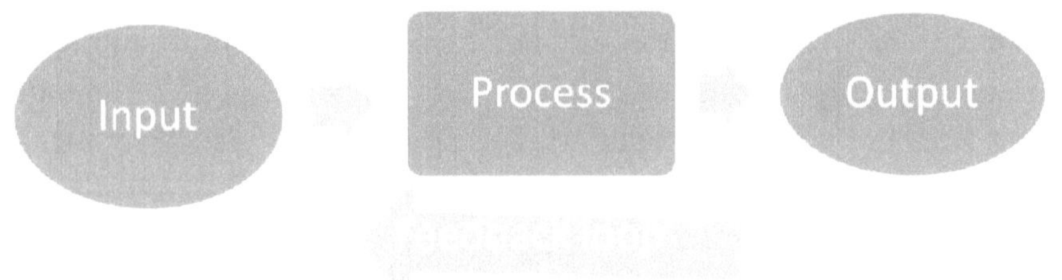

Fig.1.2 Input Output Process of a System

1.2 Design, Engineering Design and Engineering?

Commonly we are used to saying "that his house or his bike is nicely designed" or "has a nice design" and the common interpretation is the shape, the aesthetics, the look, the color etc., we generally overlook the technical brief of function, quality, reliability, durability etc.,

Design as a definition seem different from various perspectives. Generally when we speak of Design the formal description is of "*Industrial Design*" meaning 'providing form'. Whereas plain English term 'design' relates to engineering and making stuff work in a certain way, i.e. designed for a purpose.

"To conceive idea for some artefact or system and/ or to express the idea in an embeddable form" (Roozonbagskels 1995).

Until 15th century "design" was no different from doing things. "Design is the first signal of human Intentions" Think about it took us 5000 years to put wheels on the baggage! Designing is no more as simple.

A comparison between industrial product design and other field of engineering can be made through six measures namely of types of problem, types of objects, form function relation, decomposition potential, language complexity and design methods.

Design is both a verb and a noun "to concise and plan out in mind to have as a purpose, intend to dense for a specific function or end (Merriam Webster) verb."

Noun design "may something is made, picture of something form and structure derivative pattern, process of designing scheme, something planned.

Later word design is to specify.

Papanek 1984 "All men are designers All that we do almost all the time is design for design is base to all human activities, ... Design is comparing an epic poem, writing a concerto, But Designing is also cleaning and reorganising a desk drawer, pulling an impacted tooth, making an apple ... educating a child.

A clear general understanding at the comprehensive domain level across the full domain and its field and outfield work with issues in all areas within the domain.

Engineering design pertains to the manner in which things are engineered; that is the process of designing product with the help of engineering tools and methods. We must understand that it is a process and we must focus on understanding the process here. As engineers we use systematic methods to create, design, analyze and evaluate to take decisions about the product and at times for the business. Engineering is interpreted differently as tools and methods that are used when designing to analyze the design behavior and properties in the real world.

Though very difficult to put down the real definition of the word; we have given some well accepted definition for your study. And all these terms are heavily related to each other and may mean differently according to the field of engineering; for part may mean differently in mechanical and differently in software engineering; so is the case with interface, relationship. Function and requirements etc.

Why are we called engineers? Does it follow from the word engine?But steam engine came into being only in 18th century whereas engineering as a profession came into existence centuries before. The word engineer has nothing to do with engine. The word engineer is derived from the Latin words ingeniare ("to contrive, devise") and ingenium ("cleverness").The work of engineers forms the link between scientific discoveries and their subsequent applications to human needs and quality of life.

So let us be ingenious and solve societal and human problems and make life easy for all!

1.3 An Introduction to Modern Engineering Design

What is your answer when asked why are you planning to be an engineer? This question may have you awestruck you for once; presumably not anymore time. One of the possible answer is, "I want to become an engineer to design...." It might as well be in your intention to design clean drinking water for rural area, a new spacecraft for ISRO, the tallest building in the world, an efficient and non-polluting automobiles, new and improved transport network, or maybe even synthetic blood. The key here is that engineers design devices, systems, or processes for the mankind.

So how do we explain engineering design? Put briefly solutions to mankind needs/problems are solved by a method known as Engineering design. Design is the soul essence of engineering. The design process is applied to problems (needs) of varying complexity. For example, mechanical engineers will apply the design process to develop an effective, smooth ride suspension system; electrical engineers will apply the process to design lightweight, compact wireless communication devices for the same vehicle; the electronics engineer would take care of steering and motion control of the vehicle and computer engineers will apply the process to design efficient navigation and routing through GPS enabled Apps.

The vast majority of complex problems in today's high technology area as in space voyages, complex tunnels and seismic bridges or warfare technologies do not depend on a single discipline of engineering for solutions; rather, they depend on teams of engineers, scientists, environmentalists, economists, sociologists, legal, personnel, and others. Solutions are dependent not only on the appropriate applications of technology but also on public sentiment as enforced through government regulations and political influence from time to time. As engineers we are empowered with the technical expertise to develop new and improved products and systems; however, at the same time we must be increasingly aware of the societal and environmental impact general and work conscientiously toward the best solution in view of all relevant factors.

1.4 The systematic design process can be conveniently represented by the six steps

1. Well define the problem on hand.

2. Gather and assemble relevant data.

3. Find out solution constraints and criteria.

4. Generate alternative solutions.

5. Decide on a solution based on the analysis of alternatives.

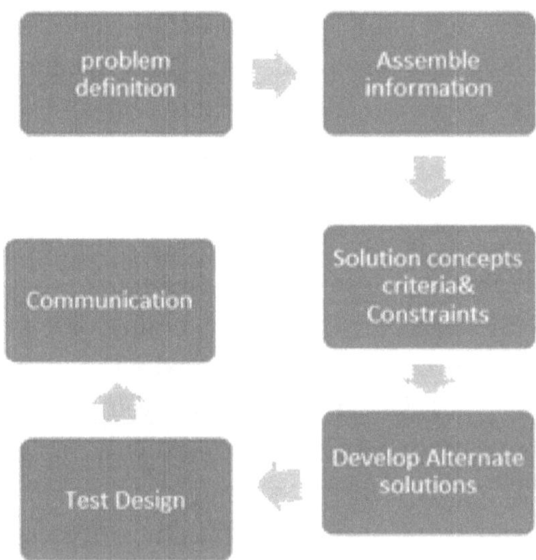

Fig.1.3Key steps in design process

6. Communicate the results.

1.5 Various Views (definitions) on Design

- Finding the right physical component of a physical system [C. Alexander]
- A goal directed problem solving activity [L.B. Archer]
- Decision making in the face of uncertainty with high penalties for error [M. Asimov]
- Relating product with situation to give satisfaction [S. Gregory]
- The performing of a very complicated act of faith [J. C. Jones]
- Design is to pull together something new or arrange existing things in a new way to satisfy a recognized need of society [G.E.Dieter]

The ABET statement on design reads as follows Accreditation Board for Engineering and Technology (ABET):

Engineering design is the process of devising a system, component, or process to meet desired needs (Fig.1). It is a decision-making process (often iterative), in which the basic sciences, mathematics, and engineering sciences are applied to convert resources optimally to meet a stated objective.

1.6 How is maths involved in design?

- can help quantify problem
- setting a budget
- Evaluate mathematical soundness of design using geometry, algebra, probability and statistics.
- Numerically Quantify performance on texts
- Calculate Probability of failure
- Revising Design; Evaluate mathematical soundness of Design using geometry, algebra, probability and statistics.
- Pugh-matrixes, Quality Function Deployment (QFD), TRIZ, Dot-voting and others

1.7 Decision making in design

The tilted square block (or rectangular block) in the design stage process represents decision making and objectively making decision is key to entire design process. Decisions are based on the fact, gut feeling, intuition or just plain thinking.

When we are certain about a decision then the decision is said to be based on facts. As an example when deciding between a number of different concepts that are all solutions to our problem some are "better" or more "efficient" etc than others based on an individual criteria. It means you have a criteria that you (say ease of assembly) set up the needs that would be analyzed on that concept and you choose the concept that has the highest value or measure (as in Pugh's matrix). It is important that these measures are quantitative, that is, you can count it without having to guess the measure. Usually quantitative decision making is done in the late stages of product development or the design process; often because we lack enough information at the initial stages about our design to state any important measure and that would then let us compare concepts. As the design progresses and becomes more and more detailed it is necessary and important not to base decisions on gut feel but purely on fact. This is much harder to realize but very important.

"I'll try and make an example of the difference between gut-feel decision and qualitative ones. Say that you make three different concepts up that you intuitively see solves your problem. However which is the best or least bad? They all solve the problem? Well here comes the systematic part into consideration. In order for you to evaluate the different concepts you have to decide criteria's that can be evaluated against each other. For example you can evaluate the concepts by the number of operations that's needed in order to produce it, and then you can evaluate the manufacturing cost for that part as well since you know what operations and what machines that are needed. You can also evaluate it in terms of logistics and transportation. Which concept has the most parts that can be transported inside a standard shipping container, are they to big or inhibit a strange geometry that makes them hard to stack or package? These things add to the transportation cost. How can the different designs fit or mate to other parts in the product or assembly? Is there any mechanical properties that are needed for a certain design? Can we exchange material to another quality or type?"

This purposeful thinking is obviously called Design for X, where X can be anything from Manufacturing, Assembly, disassembly, exchangeability,

Maintenance (refer P. for more details)Here consideration is also given to quantity in which the product is going to be manufactured if the product will be manufactured in large quantities where little changes in the design may mean big difference in the price for the end product. Also the quantity produced affects the time that you can put into the design phase if you are to make some profit out of it.

For taking decisions numerous tools like Pugh-matrixes, Quality Function Deployment (QFD), TRIZ (theory of the resolution of invention-related tasks in Russian), Dot-voting and others. For examining the part we can adapt and adjust to our specifications. Using engineering methods and modern analysis tools to act as a compliment to traditional manual tools and methods you can design, analyse and redesign your product in a quick and efficient manner.

1.8 How is Computer and Software involved?

What is CAD (COMPUTER AIDED DESIGN)?

CAD is more or less related to creating 2D and 3D virtual models of your product. Increasing computational capability has made it easier for us to use 3D models advantageously. With 3D models it is much easier evaluate and design the shape of the parts and it's interaction with other components, since now we can move, rotate and zoom into the design. 3D models has another big advantage is that can be re-used in different ways when analyzing the physical properties of the parts and their interaction as well as the possibility to adapt the manufacturing and production facilities completely within the computer (CAM Computer Aided Manufacturing). This type of thinking is known as Master Modeling where you have only master design that we make small adjustments on when analyzing. Modern tools used to analyze the product is mentioned below.

Finite Element Analysis (FEA)

- Primarily used in structural analysis of mechanical components but now the method is so general and used as well in thermodynamic and other physical analysis. It is quite easy to grasp how it works and functions for easier but demands quite a lot of understanding to master.

- Dynamics Analysis (Multi-body)
  - Mainly the dynamic simulation of mechanical systems such as cars, trains, engines, etc. Basically you create assemblies of parts (CAD-models) that you connect together using different joint such as revolute joints etc. you add material properties to the parts and then you run a simulation during a set time-frame. The software calculates forces, momentum and other interesting properties that may be hard to calculate by hand. Often the results from these kind of simulations can be used as input to Finite Element Analysis.

- Computational Fluid Dynamics (CFD)
  - Basically used to analyze different fluid and their interaction with components etc. Different modeling techniques exist for the study of fluids and their dynamics.

- Thermodynamics
  - Programs that helps us to analyze thermal effects on components and systems within our product. Different modeling techniques exist for their study and analysis.

- Electric schematics
  - Electronic CAD-software lets us not only create the schematics for the electronics it also can help us analyze the electronic system and test it out virtually before any physical prototype is built. It can generate circuit board design as well as 3D-modells of the components and circuit board.

- Mechatronics
  - Lets us create a electro-mechanical system of both rigid parts as well as electric or other components. It enables you to design how the sensors and controls should behave and examine that the mechanical system works the way you want.

- Production cells
  - If our product design is in 3D we may incorporate its manufacturing and assembly into the production line. Where you can examine ergonomics of manufacturing and also design robot cells etc.

Computer Aided Engineering (CAE) or CAx is an umbrella term for these kind of tools and analysis methods. Many of these tools requires good engineering and mathematical knowledge in order to utilize them to it's fullest however some of them are becoming more and more convenient and easy to use. Now these tools has become quite common so that we can download and use trial versions of the software. Often universities and higher education have student licenses so that you can use the software within specific courses. Also quite a few of good software's are available for free if you just want to learn basic skills.

Also some of the bigger CAE-systems that are commercially available have many of these functionalities built within them or can be added with modules for example;

Windmill software for PLM

Autodesk Revitstructure, AutoCAD for modeling and analysis.

Creo, ANSYS, CATIA, Unigraphics –for modeling and (FEA)Finite Element Analysis.

1.9 The Process of Design

Putting it simply design is "a structured problem-solving activity." A process, as we know is a phenomenon identified by step-by-step changes leading us towards a required result. Both these definitions suggest the idea of an orderly, systematic approach to a desired end. The design process, however, in practice is not linear. That is, one does not necessarily achieve the best solution by simply proceeding from one step in the process to the next. New discoveries, additional data, and previous experience with similar problems generally will result in several iterations through some or all the steps of the process (Fig.).But does this linear process work in the case of International Space Station for a lift or for a building or a simplistic can opener needs much to be looked upon. Or do we need a new customer focused non- linear iterative process is the igniting question.

The design process was seen as an isolated thing in the past where we first design our product and then when that was finished we transfer the drawing to the production facilities where the production engineers decided how to produce it. The design process was considered as a linear process where all the tasks where seen as separate tasks that was dependent on the task that was performed in an earlier step. The design was "***thrown over the wall***" (Figure 1 top part) as thing that when finished, it to some other task and it was not our problem any more. Today it is quite obvious that this was not a good practice since errors that were discovered in the design at later stages in the process were very costly since a lot of things had to be redone. In order to cope with these problems and try to avoid costly late design changes a Concurrent Engineering design process (Figure UNIT 4 P.) have been used extensively.

It is important to recognize that any project will have time constraints. Normally before a project is approved a time schedule and a budget will be approved by management.

Design process is special in that multiple task are examined and designed concurrently, for example engineers from the design, analysis, manufacture and logistic tasks may work in a team in order to design the best possible product that not only suit the customer needs but as well suits the internal company needs as well as packaging and logistics. Using these kind of processes usually involves using computer aided tools and methods that may be quite complex and advanced however some of the methods are leaning more towards new ways of thinking and work methods that can be adopted with minor investments. Important advantages in using concurrent methods is that you decrease the time it takes to bring the product to the market. Given below are a couple of diagrams showing the difference between traditional "over the wall" product development and using Concurrent Engineering methods.

Fig.1.4 General product development (design) process

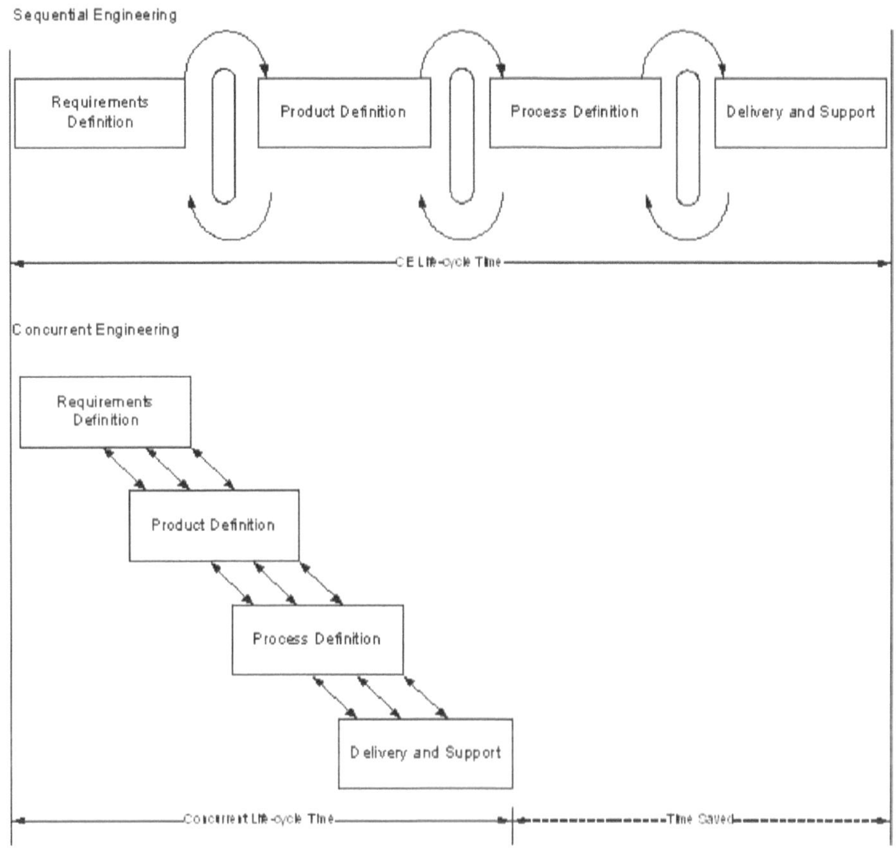

Fig.1.4 Concurrent engineering process.

The key difference between the sequential traditional model and the Concurrent Engineering one is that product development is seen as an iterative approach where you can go back and forth and redo things until it is satisfactory before you build something instead of just blindly going forward until you can't go any further due to some complicated problem.

For your initial introduction to the design process, we will explain in more detail what is involved at each of the six steps above. Simply memorizing the steps will not give you the needed understanding of design. We suggest that you take one or more of the suggested design problems at the end of the chapter, organize a team of two to four students, and develop a workable solution or each problem selected. By working as a team you will generate more and better solution ideas and develop a deeper understanding of the process. The process begins with a definition of the problem (Step 1) to be solved.

Fig.1.5 The International Space Station as was seen by the departing Space Shuttle *Atlantis* during STS-132.

The International Space Station (ISS) is currently the largest artificial body, a space station, or a habitable artificial satellite, in low Earth orbit and can often be seen with the naked eye from Earth. First component of ISS was launched into orbit in 1998, and the ISS is now comprised of pressurised modules, external trusses, solar arrays, and other components.ISS components were launched by Russian Proton and Soyuz rockets as well as American Space Shuttles. The ISS serves as a microgravity and space environment research laboratory in which crew members conduct experiments in physics, astronomy, meteorology, biology, human biology and other fields. ISS is best suited for the testing of spacecraft systems and equipment required for missions to the Moon and Mars. The ISS maintains an orbit with an altitude of between 330 and 435 km (205 and 270 mi) by means of re boost manoeuvres using the engines of the Zvezda module or visiting spacecraft. It completes 15.54 orbits per day.ISS is the ninth space station to be inhabited by crews, following the Soviet and later Russian Salyut, Almaz, and Mir stations as well as Skylab from the US. ISS is continuously occupied for the last 16 years and 152 days since the arrival of Expedition 1 on 2 November 2000. This is the longest continuous human presence in space, surpassing the previous record of 9 years and 357 days held by Mir. The station is serviced by a variety of visiting spacecraft: the Automated Transfer Vehicle, the H-II Transfer Vehicle, Dragon, Soyuz,

Progress and Cygnus. It has been visited by astronauts, cosmonauts and space tourists from 17 different nations.

Briefly exploring possibilities around us let master this art by an extensive 10-Step process

1.10 The 10 steps of Design

Research steps (1–6)

1. Identify Needs
- What's the problem?

Underlying problem the customer wants to solve – for example people didn't use toothbrush for brushing their teeth until 1938 they used other arrangements for cleaning teeth like *neem* and *jamun* sticks were chewed and used as brush ultimately led to toothbrushes and using several tons of nylon for that purpose.

2. Information phase
- What exists?
3. Stakeholder phase
- what's wanted and who wants it?

Who are the stakeholders for this books you, your parents because they purchase, publishers for their interest in publishing. Authors for their penchant to write, Teachers using this books for reference, educational quality if this turns out to be good and useful book and so on a very broad list has to be thought out..

4. Planning/operational Research
- What's realistic? What limit us?
5. Hazard Analysis
- What's safe (what can go wrong?)
6. Specifications
- What's required?

Design Steps (7-9)

7. Creative Design
- Ideation

8. Conceptual Design
- Potential solution
- Is it logical? Does it make sense? Can it transform to product functions psychologically and physically?

What does VCR do?-It has two heads and plays, stop, pause, record, eject mechanism. Is it still useful if we for the sake of argument remove recording or for that matter play head? What if my business is video duplication; now if i remove the play head and keep the play head –it is convincingly a good design my need is solved at a lesser cost and in least time.

9. Prototype Design
- Create a reason of the preferred Design
- Doesn't work? If not redesign
10. Testing
- Does it work? if not redesign

1.11 Examples of Designing Practice

Design- Steps to cooking dinner

Identify needs – hungry, stay alive, social occasion, to earn money, to like

Information phase- cook books, Sanjeev Kapoor Khazana, youtube

Stake holder phase- neighbour, cops, your parents, friends

Planned Research- time, cost

Hazard analysis- burns, fire alarm

Specification-taste it before serving!

End Solution

Design-A surprise birthday dinner for your best friend

Identify Needs- awesome, surprise, reciprocity effect

Information Phase-their hobbies & likes, how big the party should be, past experience, google

Stakeholder Phase-neighbours, cops, your friends

Planned Research-time, cost

Hazard analysis- friend finds out ahead of time taking away the surprise element, cops, weather

Engineering

Design- Making a car fuelled by nuclear reactor

Identify Needs-Different fuel, Cleaner energy, Faster car

2Information phase- (submarine fuelled by nuclear reactor) small module nuclear reactor, talk to professor, regulations.

Stakeholder phase- Nuclear Engineer, government, oil companies, car manufacturer, mechanics, designer, consumer

Planned research- Fusion/Fission, spent fuel disposal, to have a enrich uranium, art making nuclear product, Would you pay for a car that never requires a fuel in your lifetime.

Hazard analysis-what happens to nuclear reactor if a car has to crash, terrorist can use, exposure to radiation, waste DIY

Work & effort required in Design - Looking at the relationships, going again & again (iterative) looking at what is wanted and what is needed & what is the context of it, engineering as you understand is no different from the description given below:

- break the problem in to parts,

- looking at relationship between those parts and optimizing the system (system like the system to generate electricity, to build a bridge etc.) and point the nos. 1, 2, 3, 4, 5 we know how to count and so sort of leads into thinking you do 1 then you do 2...

- Not the way if you recognize that actually in good design you are doing it in parallel you are at no. 5 doing the hazard analysis you realise some of the assumptions you made early is not going to work that may because they are dangerous to you or nature you go back to step 1 ;so this fast a very good linear way of representing it but in actual the implementation you are going to move in and out of that stage. A design process involves interaction with a client to develop a product, method, or system that meets the needs of a user.

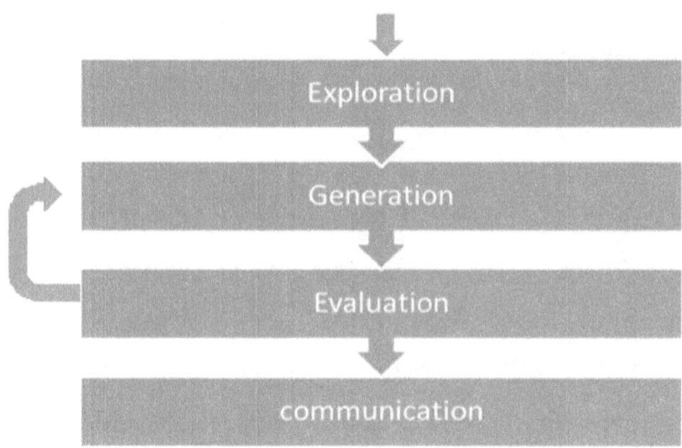

Fig.1.6. Four stage design process (Nigel Cross)

For an engineering perspective, Cross developed this "simple descriptive model of the design process, based on the essential activities that a designer has to perform. It ends with the process as the communication of a design, ready for manufacture. Before this, the design proposal is subject to evaluation against the goals, constraints and criteria of the design needs. The proposal itself arose from the generation of a concept by the designer, usually after some initial exploration of the ill-defined problem space."

Cross's model includes communication in the final stage. Though Archer (1963) was the first to include communication as a stage in the design process model.

1.12 Engineering Example:- Designing a building

1. What is the problem?
- How do we fit many people in limited space?
- How do we provide shelter?
2. Possible solution
- Building
- underground shelter
3. Design of options once we decide on building
- Material options
- Dimensions of Building

(following will affect stability of Building)

- How many floors?
- Window Doors/ventilation system etc

4. Testing Design
- Test under various loads
- Test under various weather conditions
- Test for natural disasters (Hurricane, earthquake, lightening room etc)

5. Revising Design
- using test data, evaluate how we can change design to improve quality.

6. Final Report
- Detailed drawings and specifications to begin construction

Design is an iterative process involving definition of needs, brainstorming possible solutions, exploring and testing alternative approaches, identifying optimal solutions, implementing selected designs, and documenting and communicating elements of the design and the process to a multiple audience, as appropriate.

Design is an iterative process in which various alternatives are morphed into a final process or product within specified constraints. Often the alternatives are derived from optimizing with respect to various resources or constraints.

1.13 What is the design process?

... a systematic development of a process or system to meet a set of customer specifications.

....using past experience or knowledge to synthesize a process or system to meet specified needs

....procedure for assembling available resources into a desired goal or product

The design process is a systematic approach for creating a solution that addresses a need. The design process, while systematic, is creative and iterative. This involved clear identification of need, criteria for success, identifying the key issues, generating, analyzing, and evaluating the solution paths, implementing and assessing the solution for continuous improvement.

the analysis of the problem is a small but important part of the over-all process. The output is a statement of the problem, and this can have three elements:

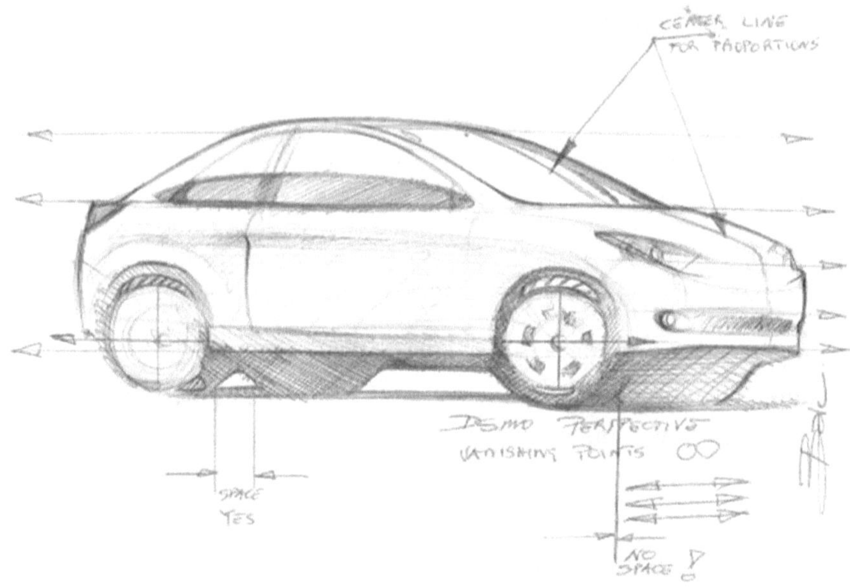

Fig.1.7 The illustration is an examples from the design of a Car early sketcher and learner.

These activities are typical of conventional engineering design. Figures below illustrate the type of design that goes on in each stage.

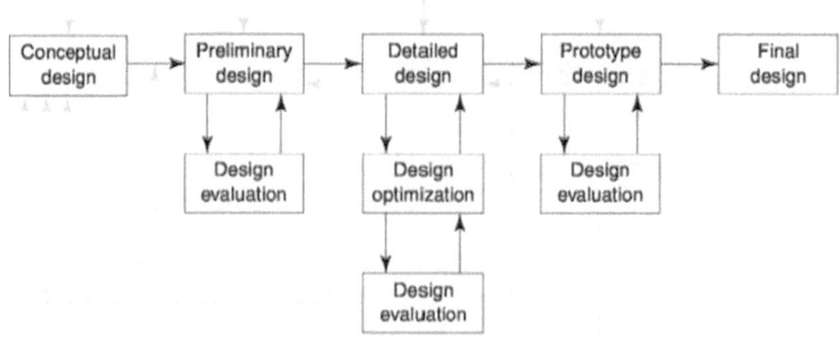

Fig. 1.8 A flow diagram for the categories of engineering design.

- Conceptual design:
  - a proper statement of the design problem
  - constraints placed upon the solution (e.g. codes of practice, statutory requirements, customers' standards, date of completion, etc.)
  - the criterion of excellence to be worked to.
  - establish function structures;
  - search for suitable solution principles;
  - combine into concept variants.
- Embodiment design:
  - starting from the concept, the designer deter-mines the layout and forms
  - and develops a technical product or system according totechnical and economic considerations.
- Detail design:
  - arrangement, form,
  - dimensions and surface properties of all the individual parts finally laid down;
  - materials speci¬fied;
  - technical and economic feasibility re-checked;
  - all drawings and other production documents produced.
  - making ideas tangible always facilitates communication

NB:The technical and process capabilities of several process used widely in the industry is given unit 3 furthering it you can cover more concepts by DIY projects, Case studies and also in the sample answers. We have taken adequate care not to demotivate you with overburdening theory and try to enlighten you with the refreshing your as you carry on with the book, A few reading of the entire book for completely grasping the subject and becoming a creative engineer or a renowned local designer.

1.14 Types of Design

Can be proposed on the basis

1. Purpose of Design-Industrial Design, System Design, Process Design

On the basis of

2. Approach to the design-Solution focused, ProblemFocused, Humancentered Design

On the basis of

3. stage of design-Conceptual Design, Embodiement Design, Detail design

On the basis of

4. Classification of product design-Design for Assembly, Design for Manufacturing, Design for Environment

On the basis of

5. Main field of Design-Software Design, ElectricalDesign, Fashion Design.

On the basis of

6. Final Communication-layout design, flowchart, CAD Drawing etc.,

On the basis of

7. Customer viewpoint-Aesthetic design, user friendly design.

And the list is endless as you can see from the figs & figs below several combinations and mixes are possible.

On the basis of

8. Time and Instant-Concurrent Designing, Reverse Engineering.

1.15 Design type and the Customer focus

*HCD (Human centred design): Here maximum focus is on the customer need.

TABLE 1.1	The Role of HCD* and Design Specializations
Experience design	
Industrial design	These are areas of focus
Interaction design	
Human-centered design	The process which ensures that the designs match the needs and capa-bilities of the people for whom they are intended.

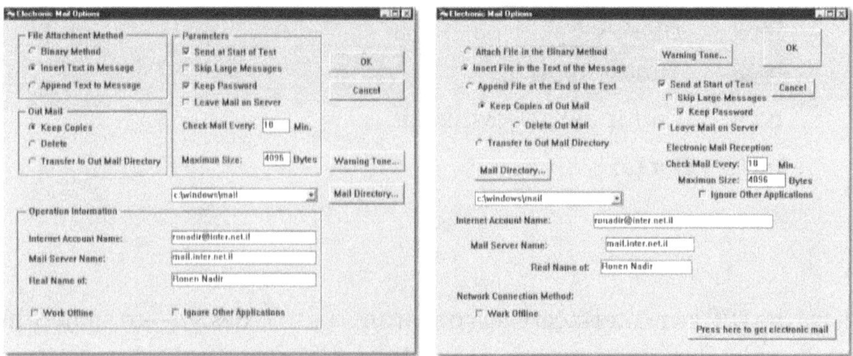

Good Design vs. Bad Design example.

Fig.1.9.Left Interface is a good design and the right one is bad interface

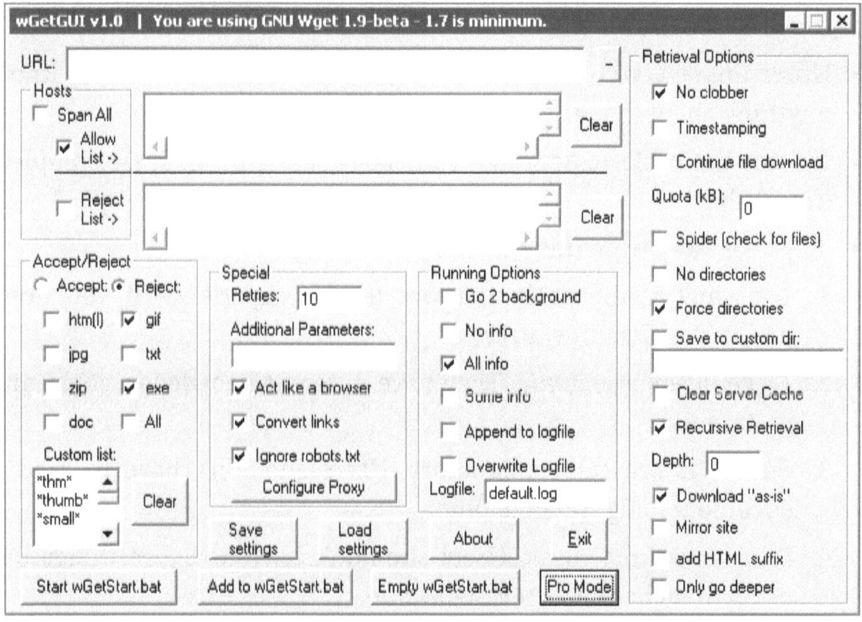

Fig.1.10 Example of bad GUI(Graphic User Interface)

we chose wGet GUI (Graphic User Interface) example because it very well illustrate some of the problems: Note that almost all the controls are text boxes, check boxes, and command buttons, which map directly to strings, integers, Booleans, and function calls. You can just see the underlying data structures that need to be filled out: struct RetrevalOptions {

```
bool      NoClobber;
bool      Timestamping;
bool      ContinueFileDownload;
int       Quota:

...

};
```

This is the UI(User Interface) everyone wants us to redesign—to something that doesn't look like it was designed by a programmer. There are some problems with this task though:

- A good design requires applying a process, not jumping to a solution.
- A good design requires understanding the target users and their goals, top tasks, environment, etc.
- Unfortunately, you are not a target user and hence don't know anything about the target users and their goals, etc..
- It's not desirable to redesign somebody else's UI without complete knowledge.
- Key take away from this exercise:
  1. You can't significantly improve a user experience if you don't understand its target users.
  2. Target users aren't you. They have different knowledge, goals, and preferences.
  3. Good UIs are self explanatory. Users shouldn't have to read a document to figure them out.
  4. Don't make me think...Users shouldn't have to figure things out using thought and experimentation.
  5. Good UIs are task centric, not feature or technology centric.
  6. The UI drives the code, not the other way around.

1.16 How to do a better linear (waterfall) design or a lateral (Vee) method

S.No	Linear Approach (Waterfall method)	
	Stages in the design process	Better Method
1	Objectives Clarification	Objectives Tree
2	Functional Statements	Function Analysis
3	Requirements Setting	Performance Specification
4	Characteristics Determination	Quality Function Deployment
5	Generating Alternatives	Morphological Chart
6	Evaluating Alternatives	Weighted Objectives
7	Improving details	Value Engineering
	Lateral Approach (System engineering Vee approach)	
1	Divergent problem explanation	Morphological Chart Brainstorming
2	Problem Structuring	Objective TTree
3	Convergence of solution	Performance specification; Synectics(a problem-solving technique which seeks to promote creative thinking, typically among small groups of people of diverse expertise)

TABLE 1.2 Better ways to do a design process

1.17 Differences in Designer Vs. Scientific Method

SUCCESSFUL DESIGNERS

- Requirements clarification by asking a set of related questions which focused on the problem structure

- Actively searched information and critically checked given requirements.
- Problem formulation summarised on the into requirements and partially prioritised them
- They held on to first solution ideas did not suppress them but returned to clarifying the problem rather than pursuing initial solution concepts in depth
- Detached themselves during conceptual design stages from fixation on early solution concepts
- Produced variant but limited the product.........and kept an overview by periodically assessing and evaluating in order to reduce the number of possible variants
- Dual exploration of problem space and solution space
- Designing skilled behavior of controlled practice and development techniques

Skilled performance -Mastery of technique and procedure

Scientist	Designers
Analysis	Synthesis
.... Found the problem use rules, optimal solution	Initial exploration and then suggest a variety of solution until a good one is found
Problem focused strategies	Solution focused strategies

1.18 Problem oriented approach vs Solution-oriented approach

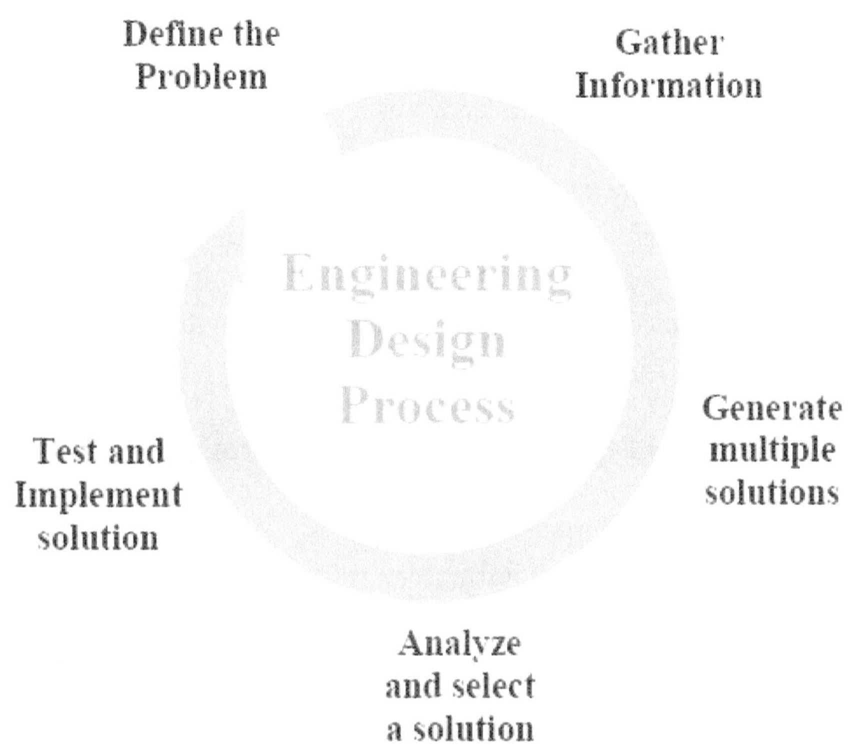

Fig.1.11 Problem Solving Methodology.

Most widely literature defines design models classically into following two categories based on the strategy used to reach the design goal

Solution-oriented: An initial solution is put forward then analyzed and modified repeatedly in the end. Herein simultaneous exploration of the design space and requirements are done.

Problem-oriented: Herein the more emphasis is on treatment and thorough understanding of the structure of problem before generating a number of possible solutions.

Lawson (1980) Strategy of putting forward multiple solutions for design is used by students who were trained to be in architectural background. Designers prefer to be creative by using trial solutions and minimizing the error (Solution oriented approach) while the scientifically trained focused more on unraveling the problem before they synthesizing solution. Real design problems are solution oriented rather than being problem oriented.

Here the basic thrust is defining and understanding the problem. It is far iterative process as it does not switch and move between problem space and solution as compared to solution focus iterative methods. This model is quite favored by the scientists as they are trained to be focused on problems as opposed to designers who are more active in the solution like architects and designers.

Most widely literature defines design models classically into following two categories based on the strategy used to reach the design goal

Solution-oriented: An initial solution is put forward then analyzed and modified repeatedly in the end. Herein simultaneous exploration of the both requirements and design space are done.

Problem-oriented: Herein more emphasis is placed on treatment and thorough understanding of the structure of problem before generating a number of possible solutions.

Lawson (1980) Strategy of putting forward multiple solutions for design is used by students who were trained to be in architectural background. Designers prefer to be creative by using trial solutions and minimizing the error (Solution oriented approach) while the scientifically trained focused more on unraveling the problem before they synthesizing solution. Real design problems are solution oriented rather than being problem oriented.

TABLE 1.3 Problem methodology Vs. Solution methodology
of design process

Problem oriented Model	Solution oriented Model
Constructing a model of problem domain	Constructing a model of solution domain
Model depends on the system	System depends on the model
Model is used to understand the problem	Model is used for optimization

Usually analytical, done at lower level, focused on the past, fishbone (cause &effect diagram) is a tool	Synectics (a problem-solving technique which seeks to promote creative thinking), modern approach (not focused on the past)
Domain specific abstractions	Implementation oriented abstractions

1.19 A Brief on Systems Engineering, Analysis and Synthesis

"Science of the Artificial" a book by Herbert Simon gives the parable of two watchmakers, each designs a watch with a hundred parts. "In the first design, the hundred parts are so thoroughly interrelated that the watch totally falls apart if any one is removed. The watch is expensive because it must be assembled in one breath and cannot be repaired by replacing parts. The second watch has similar performances, but its parts are grouped into ten modules, which can be replaced if defective. Because it is simple to assemble and easy to modify, the second watchmaker is able to offer his products at a lower price and drive the first watchmaker out of business."

First watch design given by Simon can be compared to a seamless web; his second design to an engineering system. Seamless web is good if built perfectly but perfection is more often than not illusory. Seamless webs as we know are prone to disasters, because they can be unraveled by the smallest of the loose ends. To control such disastrous scenarios is a major reason for modularity this way effect of a tiny flaw propagating unhindered and creating web-wide havoc is one of the root is minimized. To limit potential damages, engineers are careful to introduce seams and modular boundaries into their systems. One can have a nice example on electric power grid. Once electricity is fed into the grid, it flows automatically according to physical laws and the conditions of the entire grid. The risk of such seamless configuration is all too well known. Effects of a small mishap, such a lightning striking out a transformer, can cascade though the grid, leading to power failures in large regions.

The infamous blackouts in the U.S;1996 Northwest power blackout and the 2003 Northeast blackout each cost damages in excess of a billion dollars. Engineers tolerated the seamless web not because they deemed it superior but because they could do little about its faults. Until very recently, they had no way to switch high-voltage currents on the grid in real time. Even so, they did try their best to install gates between regions for damage control. New England escaped the 2003 Northeast blackout because the gate designed to disengage it from the grid worked in time – it was saved by a seam.

In India the system failed again at 31 july 2012 (13:02 IST) due to a relay problem near the Taj Mahal.As a result, power stations across the affected parts of India again went offline. NTPC Ltd. stopped 38% of its generation capacity.Over 600 million people (nearly half of India's population), in 22 out of 28 states in India, were without power

More than 300 intercity passenger trains and commuter lines were shut down as a result of the power outage.The worst affected zones in the wake of the power grid's collapse were Northern, North Central, East Central, and East Coast railway zones, with parts of Eastern, South Eastern and West Central railway zones. The Delhi Metro suspended service on all six lines, and had to evacuate passengers from trains that stopped mid-journey, helped by the Delhi Disaster Management Authority

We have systems operating seamlessly without being a seamless web; internet is an example which being a patchwork of many networks: landline telephone networks, satellite links,wireless computer connections and much more. All participating networks use packet switching. Otherwise each has its peculiar internal operating principles. A basic design principle of the internet is to preserve as much as possible the autonomy of these internal principles, so that each network can be individually modified and improved. To tie disparate network together in the internet, engineers design routers and protocols at the interface between two networks, so that signals can pass smoothly between them. The routers are seams, good seams. The superiority of an engineering system lies not in seamlessness but in its good seams or good interfaces between its parts.

This illusion of perfection, the gist of the seamless-web metaphor and associated "systems thinking" is the holistic aura that shields it from any kind of critical analysis. The resistance to analysis makes them look profound, but they are basically obscure and muddled, which are major critiques of postmodern studies. In stark contrast, the systematic approaches in science and engineering strive for clarity.

One must recognize the difference between seeming profound and being profound. So different is dealing with real-world complexity from using the word "complexity" to decorate simplistic ideas dreamed in arm chairs. The purpose of Simon's parable is intended to illustrate an approach for designing complex artificial systems that must operate in the real world, but the wisdom applies equally to research in natural science. In principle, the universe is a whole in which everything is connected to all others. Gravity and electromagnetism, the two forces that act between all things above the nuclear level, have infinite ranges. However, if we must treat the universe as a seamless web and grasp everything in it at once, our tiny brains would be so overwhelmed that we would fail to understand anything at all. None of our concepts would be valid, because concepts invariably make distinctions and "carve nature at its joints," as Plato said.

Science is partly successful because scientists take one step at a time, to bite off what they can chew, and refrain from confusing grandiloquent such as seamlessness. Instead of trying to tackle the whole universe at once, scientists take things apart and examine bits and pieces, acknowledging their own limitations. The methods of Socrates were said to be division and collection; Galileo, resolution and composition. Descartes and Newton discusses analysis and synthesis. Engineers practice functional decomposition and physical integration in systems design.

Analysis clarifies. Analysis is also called reduction, and "reductionism" to scientists means the importance of analysis. However, "reductionism" has also become a philosophical dogma asserting that a system is nothing but its constituents, e.g., you are nothing but your genes or neurons. Ideological reductionism, which slights synthesis, has engendered much debate in the philosophy of science, debates that generate more heat than light.

Holism rejects analysis and sees only the whole. Reductionism rejects synthesis and sees only the parts. The systems approach in engineering integrates analysis and synthesis. It brings to relief the dictionary definition of a system as a whole with interrelated parts, emphasizing internal structure

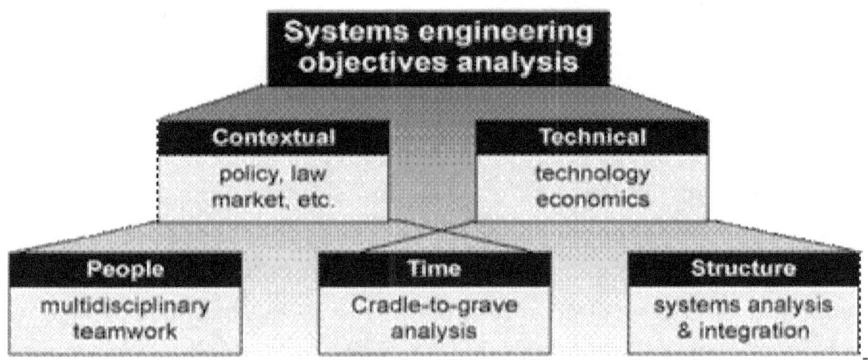

Fig.1.11 systems engineering an holistic view.

1.20 A Note on Functions, Abstraction and system modeling

Adam Smith, whose Wealth of Nations appeared in 1776 observed a relation between machines and abstract systems: "Systems in many respects resemble machines. A machine is a little system, created to perform, as well as to connect together, in reality, those different movements and effects which the artist has occasion for. A system is an imaginary machine invented to connect together in fancy those different movements and effects which are already in reality performed." Smith's remark removed the shadow on the abstraction from concrete machineries prevalent in engineering systems theories.

A part has two kinds of characteristic. The first is physical properties: its materials, structures, and motions. The second one is functional characteristics: what it is for, what services it performs. Systems theories abstract from physical properties and bring to relief functional characteristics. Now we can distinguish two broad forms of concepts. Substantive concepts address *what are*, things and their properties. Functional concepts address *what for*.

Functions tactly points to a context, the larger environment in which a system operates and provides services

Amongst functional concepts are dispositional properties, which indicate what a thing would do under certain circumstances. We know properties such as solubility and flammability occur in physics. They are much more numerous in engineering, whose cradle-to-grave (or product life cycle) considerations cover properties such as affordability, reliability, durability, manufacturability, and a host of other abilities. Many of such dispositional properties are kind of abstract and intangible. To treat them property engineers have to introduce concepts to define them clearly and performance metrics to measure them. Such concept formation is a kind of scientific thinking. we give a picture below to induce such thinking if you spare some time on it you can easily get some insight into control engineering/system engineering. If you study in such a fashion it would be easy to see how electrical engineering and mechanical are much similar; you know the analogy of voltage and potential energy similarly you will see the analogous relation between force and current and velocity between mass and inductance etc.,That every function/output could be reduced to mathematics and transforms(Laplace or Fourier for instance)

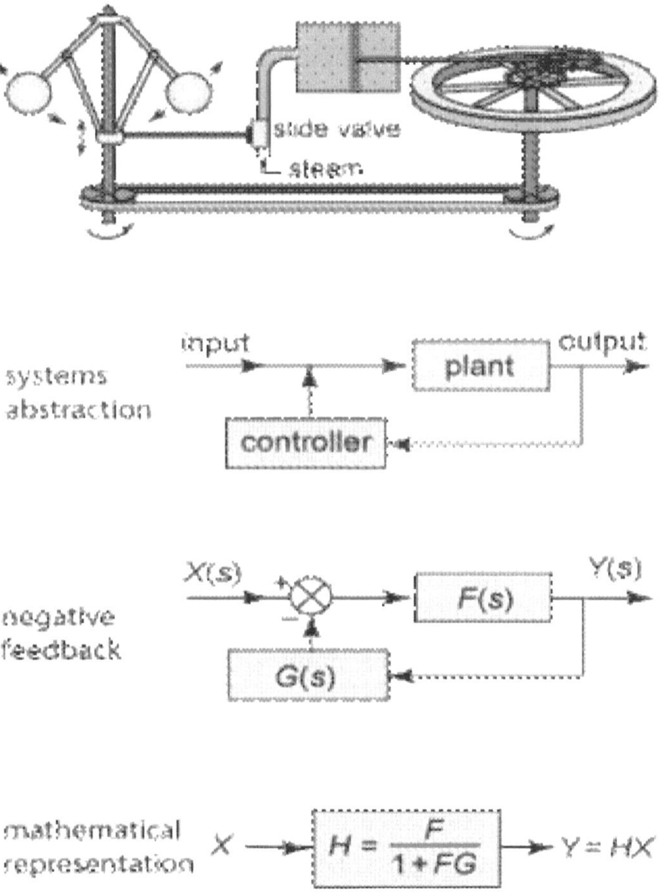

Fig.1.12 Simple illustartion of system abstraction to mathematical transformation

Functions are paramount in engineering, because engineered systems are designed for service. Thus it is no surprise that they get the center stage of systems theory.

Not surprisingly, systems abstraction is absent in physical sciences and peculiar to engineering, which aims to design systems that serve people.

Let us consider an example of signal processing, a widely applicable systems theory. A signal is something that carries information, and signal processing is the transformation of signals for efficient manipulation, transmission, or storage. Signals come in a variety of physical media: acoustic, mechanic, electromagnetic. Signal processors engage disparate physical mechanisms: electric, piezoelectric, electronic. When you speak on a phone, your information is carried in acoustic signals, which are immediately transformed into electromagnetic signals, which go through modulation, digitization, coding, multiplexing, demultiplexing, decoding, and other signal processing procedures before they are transformed back into acoustic signals heard by your friend on the other end.

Signal processing theory abstracts from most physical properties, retaining only a few essential features such as the wavelength of signals. It aims to capture, in mathematical terms, the functions of signal processors, how they transform the forms of signals, for instance the function of digitization transforms a signal from analog form to digital form. Because it captures the fundamental principles covering a wide variety of physical mechanisms, signal processing theory is applicable everywhere from telecommunications to scientific experiments.

Employing different levels of abstraction to represent complex phenomena, focusing on the topic of interest at hand, is another common method in the natural science. It is used here in engineering science.. Functional concepts occur less in physics than in biology, especially evolutionary biology, where natural selection selects adaptable functions. Functional concepts are sort of controversial in natural sciences, because they connote some sense of purpose. Evolutionary biologists who evoke them often refer to engineering; there functional concepts are intuitive, prominent, and successful.

1.21 Designing Opportunities in the world around us.

The world continues to change sometimes continuously at other times radically. As a professional engineer, you will occupy centre stage in many of these changes in the immediate near future and will become even more involved in the more distant future. The National academy of Engineering has identified 14 "Engineering Grand Challenges." These include:

1) making solar energy economical; 2) providing energy from fusion; 3) develop carbon sequestration methods; 4) manage the nitrogen cycle; 5) provide clear water; 6) restore and improve urban infrastructure; 7) advance health informatics; 8) engineer better medicines; 9)reverse-engineer the brain; 10) prevent nuclear terror; 11) secure cyberspace;12) enhance virtual reality; 13) advance personalized learning; and 14) engineer tools of scientific discovery. (Source: National Academy of Engineering of the National Academies, "Grand Challenges for Engineering,"

www.engineeringchallenges.org,.)

The huge tasks of providing solutions to these problems will challenge the technical community beyond anyone's

Imagination. Engineers of today have nearly instantaneous access to a wealth of information from technical, economic, social, and political sources. A key to the success of engineers in the future will be the ability to study and absorb the appropriate information in the time allotted for producing a design or solution to a problem. A degree in engineering is only the beginning of a lifelong period of study in order to remain informed and competent in the field Engineers of tomorrow will have even greater access to information and will use increasingly powerful computer systems to digest this information. They will work with colleagues around the world solving problems and creating new products. They will assume greater roles in making decisions that affect the use of energy, water, and other natural resources. Engineering design solution considerations for energy, the environment, infrastructure, and global competitiveness are addressed in the following sections.

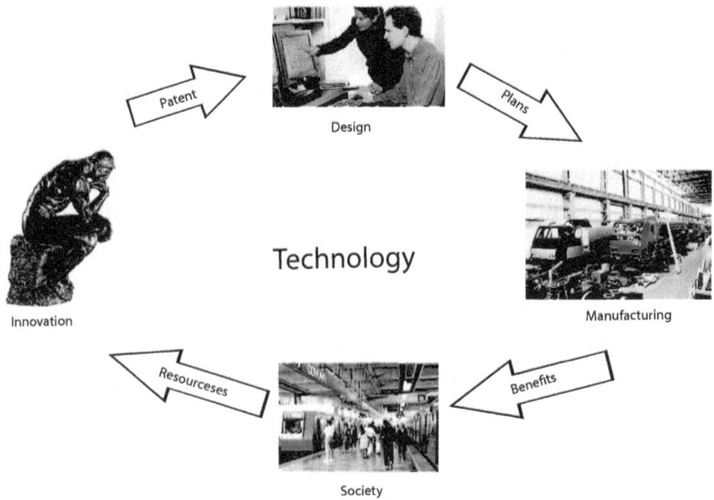

Fig.1.13 Design engineering solutions to societal needs

Have a firm belief in your Engineering Degree!

Satya Narayana Nadella chief executive officer (CEO) of Microsoft,Nadella did his schooling from the Hyderabad Public School, Begumpet before doing his B.E. in electrical engineering from the Manipal Institute of Technology in 1988 Manipal, Karnataka. (then part of Mangalore University),

Nadella went to the U.S. to study for an M.S. degree in Computer Science at the University of Wisconsin–Milwaukee receiving his degree in 1990. Later he pursued his MBA from the University of Chicago Booth School of Business.

"Nadella said he "always wanted to build things." He knew that computer science was what he wanted to pursue, but that passion was at precocious stage while studying at Manipal University. "And so it [electrical engineering] was a great way for me to go discover what turned out to become a passion," he says.

Author assumes in this famous quotes you rediscover yourself.

1.21.1 Energy

For rapid technological development as India envisages into the 21ˢᵗ century India requires vast amounts of energy. With a finite supply of our greatest energy source, fossil fuels, alternate renewable energy sources must be developed and existing sources must be controlled for efficient usage and also for reducing greenhouse gases.

A key factor in the design of products must be minimal use of energy.

As demand increases and supplies become scarcer, the cost of obtaining the energy increases and places additional burdens on already financially strapped regions and individuals. Engineers with great vision are needed to develop alternative sources of energy from the sun, radioactive materials, wind, biomaterials, and ocean and to improve the efficiency of existing energy consuming devices Ethanol and biodiesel are two fuels that can be produced from renewable resources that can assist in reducing Indian dependence on foreign sources of energy. Waste-to-energy and biomass resources are also recognized by the Government Department of Energy as renewable energy source and are included in the department's tracking of progress toward achieving the government's renewable energy goal. (As an example Buses are plied in Nagpur using Biogas see fig 1.6) Along with the production and consumption of energy come the secondary problems of pollution and global warming. Such pollutants as smog, acid rain, heavy metals, nutrients, and carbon dioxide must receive attention in order to maintain the balance of nature. Also, increasing concentrations of greenhouse gases are likely to accelerate the rate of climate change, thus causing global warming. According to the National Academy of Sciences, the Earth's surface temperature has risen by about 1 degree Fahrenheit in the past century, with accelerated warming during the past two decades.

will now be joined in the city by some 100

Fig.1.14 India's first ethanol powered bus will be now joined in the city by some 100 Scania buses running on ethanol and biogas.

1.21.2 Environment

Our insatiable demand for energy, water, and other national resources creates imbalances in nature that only time and serious conservation efforts can keep under control.The concern for environmental quality is focused on four areas: cleanup, compliance, conservation, and pollution prevention. Partnerships among industry, government, and consumers are working to establish guidelines and regulations in the gathering of raw materials, the manufacturing of consumer products, and the disposal of material at the end of its designed use.

The American Plastics Council publishes a guide titled Designing for the Environment, which describes environmental issues and initiatives affecting product design. All engineers need to be aware of these initiatives and how they apply in their particular industries:

Design for the Environment (DFE): Incorporate environmental considerations into product designs to minimize impacts on the environment. Environmentally Conscious Manufacturing (ECM) or Green Manufacturing:

Incorporating pollution prevention and toxics use reduction into product manufacturing.

Extended Product or Producer Responsibility (Manufacturer's Responsibility or Responsible Entity): Product manufacturers are responsible for taking back their products at the product's end of life and managing them according to defined environmental criteria.

Life Cycle Assessment (LCA): Quantified assessment of the environmental impacts associated with all phases of a product's life, often from the extraction of base minerals through the product's end of life.

Pollution Prevention: Prevent pollution by reducing pollution sources (e.g., through design) as opposed to addressing pollution after it is generated.

Product Life Cycle Management (PLCM): Managing the environmental impacts associated with all phases of a product's life, from inception to disposal.

Product Take back: The collection of products by manufacturer at the product 'send of life.

Toxic Use Reduction: Reduce the amount, toxicity, and number of toxic chemicals used in manufacturing.

As you can see from these initiatives, all engineers regardless of discipline must be environmentally conscious in their work. In the next few decades we will face tough decisions regarding our environment. Engineers will play a major role in making the correct decisions for our small, delicate world. The basic water cycle—from evaporation to cloud formation, then to rain, runoff, and evaporation again—is taken for granted by most people. and our water supply becomes a crucial problem. In addition, some highly populated areas have a limited water supply and must rely on water distribution systems from other areas of the country. Many formerly undeveloped agricultural regions are now productive because of irrigation systems. However, the irrigation systems deplete the underground streams of water that are needed downstream.

These problems must be solved in order for life to continue to exist as we know it. Because of the regional water distribution patterns, the state government must be a part of the decision-making process for water distribution.

One of the concerns that must be eased is the amount of time required to bring water distribution plan into effect. Government agencies and the private sector are strapped by regulations that cause delays of several years in planning and construction. Greater cooperation and a better informed public are goals that public works engineers must strive to achieve. Developing nations around the world need additional water supplies because of increasing population growth. Many of these nations do not have the necessary freshwater and must rely on desalination, a costly process. The continued need for water is a concern for leaders of the world, and engineers will be asked to create additional sources of this life-sustaining resource.

1.21.3 Infrastructure

All societies depend on an infrastructure of transportation, waste disposal, and water distribution systems for the benefit of the population.In India much of the infrastructure is in a state of deterioration without sound plans for upgrading. For example:

1. Commercial jet fleets include aircraft that are 35 to 40 years old. Major programs are now underway to extend safely the service life of these jets. In order to survive economically, airlines must balance new replacement jets with a program to keep older planes flying safely.

2. OverOne-half of the sewage treatment plants cannot satisfactorily handle the demand.

3. The interstate highway system, over 50 years old in many areas, needs major repairs throughout. Over-the-road trucking has increased wear and tear on a system designed primarily for the automobile. Local paved roads are deteriorating because of a lack of infrastructure funds.

4. Many bridges are potentially dangerous to the traffic loads on them.

5. Railroads continue to struggle with maintenance of rail beds and rolling stock in the face of stiff commercial competition from the air freight and truck transportation industries.

6. Municipal water and wastewater systems require billions of rupees in repairs and upgrades to meet public demands and stricter water-quality requirements.

It is estimated that the total value of the public works facilities is over trillions of Rupees.

To protect this investment, innovative thinking and creative funding must be fostered. Some of this is already occurring in road design and repair.

For example, a new method of recycling asphalt pavement actually produces stronger product. Engineering research is producing extended-life pavement with new additives and structural designs. New, relatively inexpensive methods of strengthening old bridges have been used successfully.

1.21.4 A Competitive Edge in the World Marketplace

We have all purchased or used products that were manufactured outside the country. Many of these products incorporate technology that was developed in the United States. In order to maintain our strong industrial base, we must develop practices and processes that enable us to compete not just with other Indian industries but with international industries. Engineers must also be able to design products that will be accepted by other cultures and work in their environments. It is therefore most important that engineers develop their global awareness and cultural adaptability competence.

The goal of any industry is to generate a profit. In today's marketplace this means creating the best product in the shortest time at a lower price than the competition. A modern design process incorporating sophisticated analysis procedures and supported by high-speed computers with graphical displays increases the capability for developing the "best" product. The concept of integrating the design and manufacturing functions shortens the design-to-market time for new products and for upgraded versions of existing products. The development of the automated factory is an exciting concept that is receiving great deal of attention from manufacturing engineers today. Remaining competitive by producing at a lesser price requires a national effort involving labour, government, and distribution factors. In any case engineers are going to have a significant role in the future of our industrial sector.

Chapter 2

Considerations of a good design

Concept generation,

Achievement of performance requirements,

Regulatory and social issues in Indian context

When you complete your study of this chapter, you will be able to:

- Understand the considerations (principles) of a good design
- Detailed concept generation with example
- What is Achievement of performance requirements
- Regulatory and social issues in Indian context
- What can go wrong with a design process? Benefits of process design as a model?
- The necessity for Prescriptive Design Method or Models.
- Problem oriented approach vs. Solution oriented approach
- Comparing French Model with Pahl&Beitz Model
- Understand Various ways to visualize design
- Designers vs, Scientist Method
- Understand problem solving Methodology

2.1 Regulatory and Social Issues in Indian context

The Government attaches much importance to the design of the built environment say for instance fsi (floor space index also fsr floor space ratio; site ratio to plot ratio). Good design is a key aspect of sustainable development (example euro 4 norm for passenger cars) is indivisible from good designing, and should contribute positively to making earth a better place to live. Along with specifications and standards play important influence on the design practice. The standards produced by such societies as BIS, ASME, Euro norms and IS represents mutual agreement among many elements (manufacturers/designers and users) of the industry. So they often represent the least common denominator standards. When good design requires more than that, it becomes necessary to raise company standards above the minimum required. Since engineers work for societal needs the code of ethics of all professional engineering societies forces the engineer to care about public health and safety. Increasingly, legislation has been passed by government to regulate several aspects of health and safety. The designer has to develop the design in such a way to prevent hazardous use of the product in an unintended but foreseeable in the immediate future. When unintended use cannot be prevented by functional design, then clear, complete, unambiguous warnings must be permanently displayed on the product and or literature (brochures, manuals etc.). The Consumer law requires the designer must take care that all of advertising materials, manuals, and operating instructions that relate to the product ensure that the contents of the material are consistent with safe operating procedures and do not promise performance characteristics that are beyond the capability of the design. Another design consideration that receives adequate attention to human factors engineering, which includes the sciences of biomechanics, ergonomics, and engineering psychology to ensure that the design can be operated efficiently by humans. In addition it applies anthropometric data and physiological understanding such design features as audio visual display of instruments and control systems. It is also concerned with capable human muscle power and response time in usage of the design.

Regulatory Issues:

The goals of regulation

- To protect consumer interests
- To increase access to technology and service
- To avoid market failure
- To foster effective competition

Creating regulatory institutions is challenging and has been a concern for all countries, especially developing and emerging countries like India, institutional capacity is weak both in qualified manpower and enforcing the standards.

TRAI for communication services

Residential building norms imposed by town and country planning, local municipal corporations

Pollution Control Boards at state levels and Ministry of Environment and Ecology

Excise laws and taxations could make the design too costly for the end user.

National transport agencies

Start Up India Stand Up India

Zero effect Zero Defect

Framework for Public Private Partnership (PPP)

Renewable Energy Corporation; Solar Energy Corporation of India-Ministry of New and Renewable Energy

Nuclear Power Corporations-DAE (Department of Atomic Energy)

Vehicle safety act and emission norms

Governmental promotion to Essential services like essential air service, low cost housing, road transport to rural areas.

Govt. is planning for a unified transport regulator, which must remain a long-term vision.

NHAI- National Highways Authority of India Act, 1998

DGCA- Directorate General of Civil Aviation - Aircraft Act of 1934, Aircraft Rules, Civil Aviation Requirements, Aeronautical Information Circulars

Airports Authority of India (merged National Airports Authority and International Airports Authority)- Airport Authority of India Act, 1994 As amended by the Amendment Act 2003

Responsible for creating, upgrading, maintaining, and managing civil aviation infrastructure both on the ground and air space in the country.

The functions of AAI are as follows: 1) Design, development, operation and maintenance of international and domestic airports and civil enclaves 2) Control and management of the Indian airspace extending beyond the territorial limits of the country, as accepted by ICAO 3) Construction, modification and management of passenger terminals 4) Development and management of cargo terminals at international and domestic airports 5) Provision of passenger facilities and information system at the passenger terminals at airports 6) Expansion and strengthening of operation area, viz. Runways, Aprons, Taxiway, etc. 7) Provision of visual aids 8) Provision of communication and navigation aids, viz. ILS, DVOR, DME, Radar, etc.

AS 9000: Aerospace basic qualification for quality management.

Ministry of Shipping- Manage the daily activities of major ports in the country- Major Ports Trust Act 1963

CEA-Central Electricity Authority

ERC electricity regulatory commission and SERC state electricity regulatory commission

Oil and Gas sector is regulated by Director General of Hydrocarbons and Petroleum and Natural Gas Regulatory Board, whereas Power has central regulator as Central Electricity Regulatory Commission and State Electricity Regulatory Commission.

Coal Mines (Nationalization) Amendment Bill, 2000

Mines and Minerals (Development and Regulation) Bill, 2011 (as introduced in the parliament) has attempted to address the key industry concerns of transparent concession systems, scientific mining, sustainable development and curbing illegal mining

Competition Commission of India (CCI)-Competition Act, 2002

Consumer Protection Act, 1986;MRTPC (Monopolies and Restrictive Trade Practices Act, 1986

Food Safety and Standards Act, 2006- Food Safety and Standards Authority of India.

BIS-Bureau of Indian Standards-BIS, Act 1968-National Building Code of India, 2005; Indian Standards Bill 2015

The main objectives of the proposed legislation (Indian Standards Bill 2015) are:-

- To establish the Bureau of Indian standards (BIS) as the National Standards Body of India.

- The Bureau to perform its functions through a governing council, which will consist of President and other members.

- To include goods, services and systems, besides articles and processes under the standardization regime.

- To enable the government to bring under the mandatory certification regime for such articles, processes or service which it considers necessary from the point of view of health, safety, environment, prevention of deceptive practices, consumer security etc. This will help consumers receive ISI certified products and will also help in prevention of import of sub-standard products.

- To allow multiple types of simplified conformity assessment schemes including self-declaration of conformity (SDOC) against any standard which will give multiple simplified options to manufacturers to adhere to standards and get a certificate of conformity, thus improving the 'ease of doing business'.

- To enable the Central Government to appoint any authority in addition to the Bureau of Indian Standards, to verify the conformity of products and services to a standard and issue certificate of conformity.

- To enable the Government to implement mandatory hallmarking of precious metals articles.

- To strengthen penal provisions for better effective compliance and enable compounding of offences for violations.

- To provide recall, including product liability of products bearing the Standard Mark, but not conforming to relevant Indian Standards.

Indian Wireless Telegraphy Act (No 17 of 1933) - WPC (Wireless Planning & Coordination Authority) Wing of the Department of Telecommunications (DoT)

RADIO SERVICES: Terrestrial radio services can be divided into two main categories: AM radio that uses medium or short wave frequency bands, and FM that uses VHF frequencies in the 88 MHz to 108 MHz band. AM radio is offered only by AIR while FM radio, which works on line-of-sight principles and can be clearly received within a local area, is offered by both AIR and private channels.

PRASAR BHARATI (BROADCASTING CORPORATION OF INDIA) ACT, 1990

Cable Television Network Rules, 1994:

INFORMATION TECHNOLOGY ACT 2000

The Communications Convergence Bill, 2000 was aimed at creating a single regulatory authority (Communications Commission of India) that would repeal the Indian Telegraph Act 1885, the Indian Wireless Telegraphy

Act 1933, the Telegraph Wire Unlawful Possession Act, 1950, and the Telecom Regulatory Authority of India Act, 1997

The Bill is applicable to the following technologies: 1. Network infrastructure facilities (e.g., earth stations, fixed links and cables, public payphone facilities, radio-communications transmitters and links, satellite hubs, towers, poles, ducts and pits used in relation with other network facilities). 2. Network services (e.g., bandwidth services, broadcasting distribution services, cellular mobile services, customer access services, mobile satellite services). 3. Application services (e.g., public cellular telephony services, IP telephony, public payphone service, Public switched data service). 4. Content application services (like satellite broadcasting, subscription broadcasting, terrestrial free-to-air TV broadcasting, terrestrial radio broadcasting). - Communications Commission of India.

DTH GUIDELINES Direct-to-Home (DTH) Broadcasting Service refers to the distribution of multi-channel TV programs in Ku Band by using a satellite system to provide TV signals direct to subscribers' premises without passing through an intermediary such as cable operator.

NBA- The News Broadcasters Association

Some Universal Perspective:

For a global market and free trade practices several universal standards and agencies ;affect our design principles and practices.

LEED® standards, in full **Leadership in Energy and Environmental Design standards**, a certification program devised in 1994 by the U.S. Green Building Council (USGBC; founded 1993) to encourage sustainable practices design and development by means of tools and criteria for performance measurement. It is "a voluntary, consensus-based, market-driven building rating system based on existing proven technology."

The five critical areas of focus, as laid out by the USGBC, are "sustainable site development, water savings, energy efficiency, materials selection, and indoor environmental quality."

1. Sustainable site development involves, whenever possible, the reuse of existing buildings and the preservation of the surrounding environment. The incorporation of earth shelters, roof gardens, and extensive planting throughout and around buildings is encouraged.

2. Water is conserved by a variety of means including the cleaning and recycling of gray (previously used) water and the installation of building-by-building catchments for rainwater. Water usage and supplies are monitored.

3. Energy efficiency can be increased in a variety of ways, for example, by orienting buildings to take full advantage of seasonal changes in the sun's position and by the use of diversified and regionally appropriate energy sources, which may—depending on geographic location—include solar, wind, geothermal, biomass, water, or natural gas.

4. The most desirable materials are those that are recycled or renewable and those that require the least energy to manufacture. They ideally are locally sourced and free from harmful chemicals. They are made of nonpolluting raw ingredients and are durable and recyclable.

5. Indoor environmental quality addresses the issues that influence how the individual feels in a space and involves such features as the sense of control over personal space, ventilation, temperature control, and the use of materials that do not emit toxic gases.

American Society of Mechanical Engineers (ASME) codes and standards

Elevators and Escalators (A17 Series), Piping and Pipelines (B31 Series), Bioprocessing Equipment (BPE), Valves Flanges, Fittings and Gaskets (B16), Nuclear Components and Processes Performance Test Codes.

Capability Maturity Models –CMM levels-level1 to level 5

A benchmark for measuring the maturity of an organization's software process.CMM defines five levels based on certain key process areas (KPA)

FDA (Food and Drug Administration)U.S, regulators have further encouraged device manufacturers to transition away from the rigid Waterfall Model (see fig. P100-101.) to the more suitable TPLC (see fig P100-101.)

Euronorms:Since year 2000 India adopted European emission and fuel regulation in stages so that we adapt to euro norm 4 by the year 2017 for four wheelers through the entire country.

International Telecommunications Union –to standardize networks say as true 4G

TL9000:Telecom quality management and measurement system standard.

2.2 Socio-Cultural Input to design

Natural Mappings Can Vary with Culture For instance left hand driven cars Almost 35% of the world population drives on the left, and the countries doing so are mainly British colonies. Japan was never a British colony, but its traffic also moves along the left. Although this habit is from the Edo period (1603-1868), it wasn't until 1872 that this unwritten rule became official. That being the year when Japan's first railway was introduced, also built with technical aid from the British. Over the year massive network of railways and tram tracks was built, and of course all trains and trams drove on the left-hand side. On 7 September 2009 the Independent State of Samoa became the third country ever to change from right- to left-hand driving. Samoa had been driving on the right since it was a German colony in the early 20th century, although it gained independence from New Zealand in 1962. Prime Minister T.S. Malielegaoi wanted to swap sides to make it easier to import cheap cars from left-hand driving Japan, Australia and New Zealand.

Fig.2.1 Samoan workers are repainting road markings

The choice dictates the proper design for interaction. Similar issues crop up in other disciplines. Very common example is the problem of scrolling the text in a computer display. This question remains debated till today should the scrolling control move the text or the window? This argument belongs to the early years of display terminals, even long before even the computers. It was decided that, that the cursor arrow keys—and then, later on, the mouse—would follow the moving window model. You can move down the window to see more text at the bottom of the screen. So in practice to see more text at the bottom of the screen you have, moved the mouse down, which moves the window and not the text.

In effect the mouse and the text move in opposite directions. With the moving text model, the mouse and the text move in the same directions: move the mouse up and the text moves up. For over two decades, everyone moved the scrollbars and mouse down in order to make the text move up.

But then smart displays with touch-operated screens have just arrived. Now it was only natural to touch the text with the fingers and move it up, down, right, or left directly: the text moved in the same direction as the fingers. The moving text model became prevalent. But as people switched back and forth between traditional computer systems that used the moving window model and touch-screen systems that used the moving text model, confusion reigned. Finally one major manufacturer of both computers and smart screens, Apple, switched everything to the moving text model, but no other company followed Apple's lead.

All systems in future move the hands or controls in the same direction as they wish the screen images to move. Predicting technology is relatively easy as compared to predicting human behavior, or as in this case, the adoption of societal/cultural conventions.

Question:

Fig.2.2 Which indicator knob is better of the two you decide. And Why?

2.3 Example of design changes as per working mother's needs.

Design and innovation, sustainability, creativity, branding, ecologically responsible design, human factors, materials, technology, graphic arts, packaging, and universal design are submitted for award every year. Approximately 40, 000 Good Design Awards  have been given since the inception of the award in 1950.In the category of babies and children, phil&teds latest Inline stroller proudly features 'Auto Stop' taking stroller safety to the next level. We have all seen the video on the YouTube, where a moments inattention can have a catastrophic outcome.

Auto Stop brake

Through our ongoing observations of stroller use we discovered that brakes are not often activated when there is no perception of danger - for example when removing shopping from the parcel tray on level ground, or answering your phone. Navigator 2.0 with Auto Stop is always safe and always convenient; Auto Stop is your silent guardian quietly activating the brake when your hands move from controlling the stroller to other things.

Auto Stop with semi-lock out function

For users who wish to push the stroller without a permanent handgrip, such as when jogging with a stroller, you can lock out the Auto Stop brake with a latch that connects to the wrist strap. If for some reason you stumble, or the stroller separates from the user more than the length of the wrist strap, the taught strap will release the latch and activate the Auto Stop brake normally.

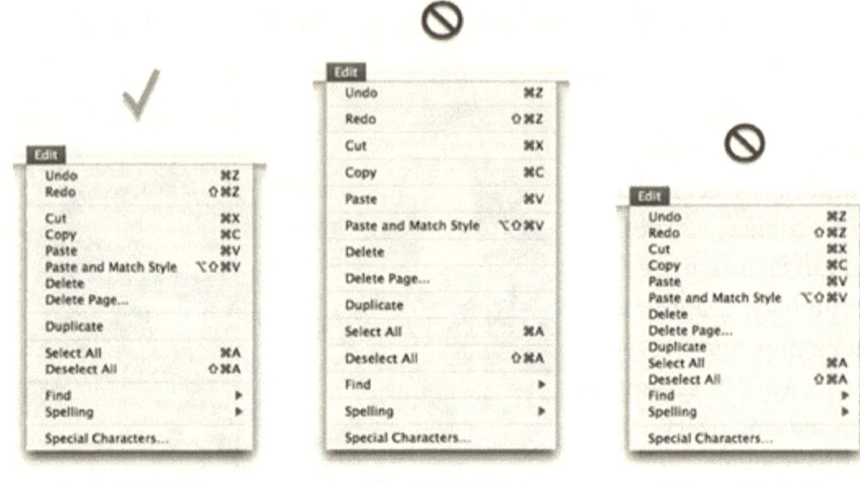

Fig.2.3 Example good vs. bad design

2.4 Considerations of a good design

1. Must be need compatible-design must be user centric
2. All the functional requirements of the product must be met.
3. Must be cost effective
4. Must be easy to manufacture
5. Must be easy to handle and assemble
6. Must be portable
7. Must be aesthetic
8. Must meet all performance requirements
9. Must be technologically relevant
10. Must be environment friendly,
11. Must be Reliable
12. Must be quality product
13. Must meet health and safety standards.
14. Must be innovative
15. Must be maintenance free –sustainable usage
16. Must be communicative to the user, must be represent able
17. Must bring excitement to the customer

18. All linkages and interfaces must be defined properly.

19. be scheduled in time.

2.5 What does a successful designer do differently?

- Clarified requirements by asking a set of related questions which focused on the problem structure.

- Actively search for information and critically checked given requirements.

- Summarized information on the problem formulation into requirements and partially prioritized them

- Did not suppress first solution ideas; they held on to them but returned to clarifying the problem rather than pursuing initial solution concepts in depth

- Detached themselves during conceptual design stages from fixation in early solution concepts.

- Produced variants but limited the product cost and kept an overview by periodically assessing & evaluating in order to reduce the number of possible variants

- Dual exploration of the 'problem space' and 'solution space'

- Designing is a skilled behavior -controlled practices and development of techniques. Skilled performance comes from mastery of technique & procedure.

Scientists Vs. Designers

Scientists	Designers
Analysis	Synthesis
Understand the problem	Internal exploration and then suggest a variety of solutions
Use Laws, find optimal solution	Find a good solution
Problem focused strategies	Solution focused strategies

A design process involves interaction with a client to develop a product, method, or system that meets the needs of a user. What we do here for example is for us to control home function say heat transfer of conduction we provide thermal insulation to our house as one of the design strategy. So here a function to mechanism to solution can easily be understood.

CONTROL FUNCTION	PHYSICAL MECHANISM	CONTROL STRATEGY
Moisture Migration	*Bulk Water*	Shedding Conveyance Drainage Storage & Drying Drain-Screen Rain-Screen Dynamic Buffer Zone 'Perfect Barrier'
	Capillary Water	Capillary Barrier Capillary Break
	Vapour Diffusion	Vapour Barrier Thermal Insulation
	Air Leakage	Air Barrier System Thermal Insulation
Heat Transfer	*Conduction* *Radiation* *Convection*	Thermal Insulation Radiation Barrier Air Barrier System
Air Leakage	*Stack, Wind and Mechanical Effects*	Air Barrier System
Solar Radiation	*Heat*	Orientation Fenestration Shading Devices Thermal Resistance Glazing Reflectance and Emissivity
	Visible Light	Orientation Fenestration Shading Devices Glazing Optical Properties

Adapted from work by Bomberg, M.T. And Brown, W.C., 1993. *Building Envelope and Environmental Control: Part 1 - Heat, Air and Moisture Interactions.* Construction Canada, 35 (1)

Fig.2.4 A control function to physical mechanism to control strategy approach.

2.6 Detailed Concept Generation With Example.

Case study-Integrated Space Mission (**MANNED SPACE VEHICLE TO THE MOON**)

MISSIONS, PROPULSION AND FLIGHT TIME

1. Explanation of Needs

I. 1. Logical Intermediate step towards the future goal of landing men on the moon and other planets now that we are capable of Manned lunar mission with lunar research.Particularly it applies to the earth reentry and recovery phases of the flight ;it is a mission which will require a considerable amount of trajectory control so rather severe requirements are needed of the navigation and control system and thereby effectively demonstrating our ability to navigate in space.This is compatible with our firmly planned booster system (Saturn in this case)

2. The second need refers to capability of lunar of earth orbital mission and that the capability of re-entry component of this vehicle of earth orbital missions in conjunction with space laboratories or space stations for research and training purposes

3. The manned advanced space vehicle should not weigh not more than 6.8 Tonnes(Metric Tons) including auxiliary propulsion and attaching structure should be designed to be compatible with the Saturn booster system and for the lunar mission

4. The vehicle should be designed for a flight time capability, without resupply for 14 days." In justification, it is to be noted that minimum time for a lunar circumnavigation mission is about 6 days. Taking a factor of safety for the total flight time may be affected by lunar perturbations, or use of no minimum time trajectories, loiter times and the desire for mission flexibility, it is believed that the vehicle should be designed for a minimum flight time of 14 days.

Now the most desirable flight path (trajectories) is to be determined from amongst several conceptual plans we make. Figure above presents a typical lunar trajectory and return. It may be used thought of the trajectory solution we would like to work with.

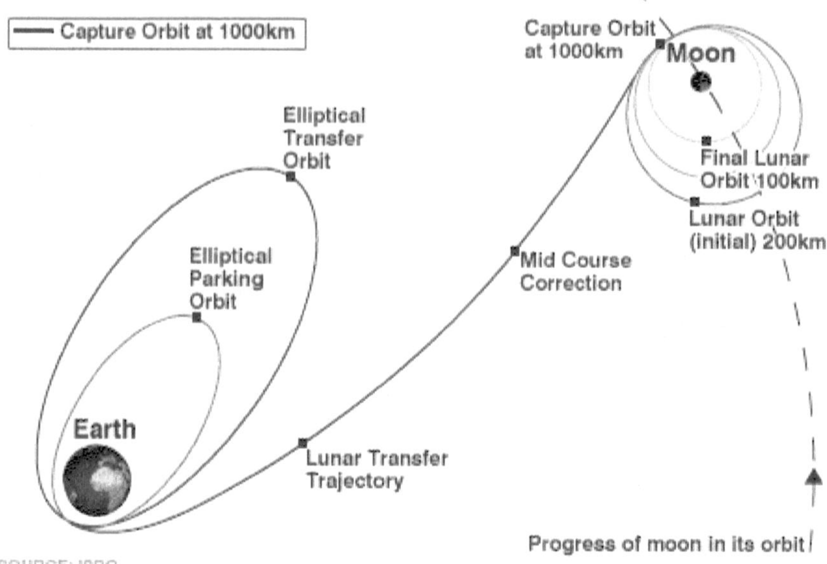

Fig.2.5 Possible lunar trajectory and return.

Now this involves a flight path which would pass the moon at a capture orbit of thousand Kms (see fig.). Two benefits will occur here: 1.We need not use large rockets 2.These trajectories which will give us so called "free" or "safe" return will be identified by study, and their compatibility with launch conditions and re-entry conditions must be well established. Considered a desirable plan for trajectory to come much closer to the moon than thousand Kms say to the order of 100 Kms first lunar orbit as in fig. would appear to be a reasonable target for survey purposes prior to man landing on the moon. To achieve this, we would propose to go into orbit about the moon. This procedure (as compared to a orbit about the moon. This path or plan of trajectory (as compared to a close pass of thousand Kms) has the advantage of allowing the decision to be made after the flight has established the integrity of the system, and also allows loitering in the neighborhood of the moon for a time period sufficient for survey purposes as long as we are in the lunar orbit.

Operational concept for landing on the moon

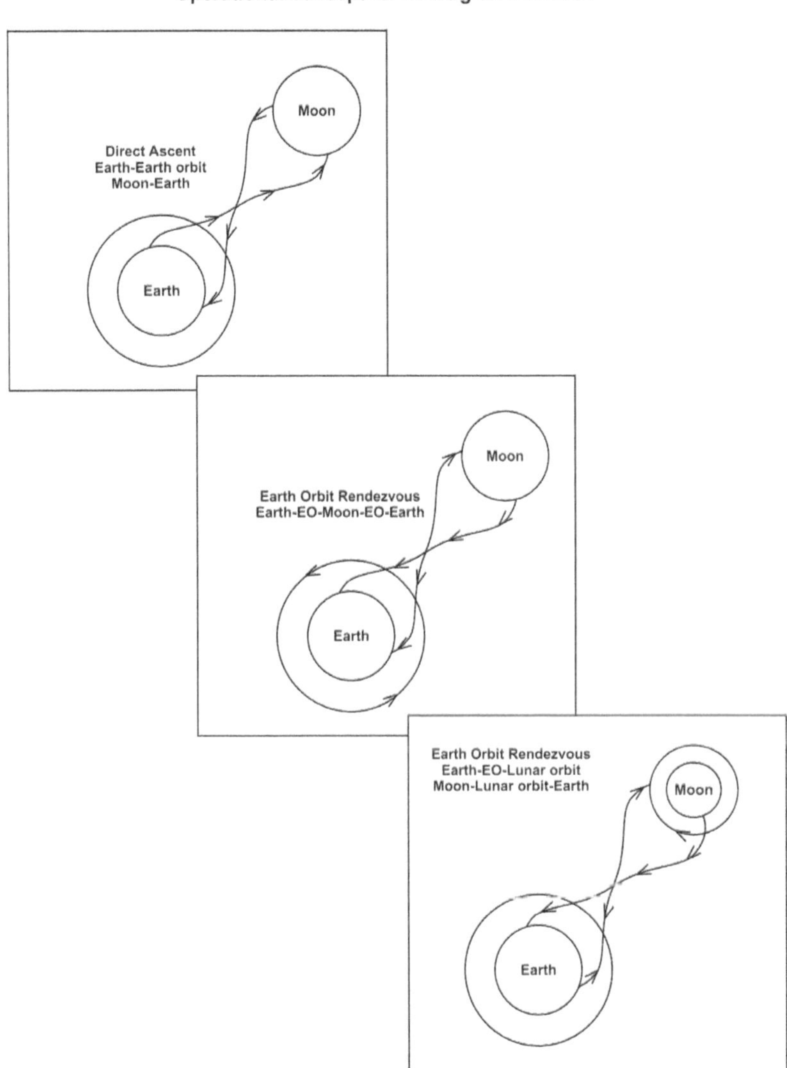

Fig. 2.6 Trajectory plans for Manned lunar (moon) space vehicles.

Many other detailed trajectory studies will naturally be required. Several examples might be various midcourse correction trajectories, elliptical transfer orbit and Lunar transfer trajectories (as in fig)and the effects of the non- uniformity of the moon's shape on the vehicle trajectory while it moves to the neighbourhood of the moon.Fig Above the selected few trajectories for manned lunar space mission as a commoner can at best conceptualize.

Earth Orbital Mission

As the second need clearly refers to capability of lunar of earth orbital mission and that the capability of re-entry component of this vehicle of earth orbital missions in conjunction with space laboratories or space stations for research and training purposes.

Hence forth the re-entry vehicle should be designed for use with both the lunar and orbiting space laboratories to avoid unnecessary duplication of vehicle development.

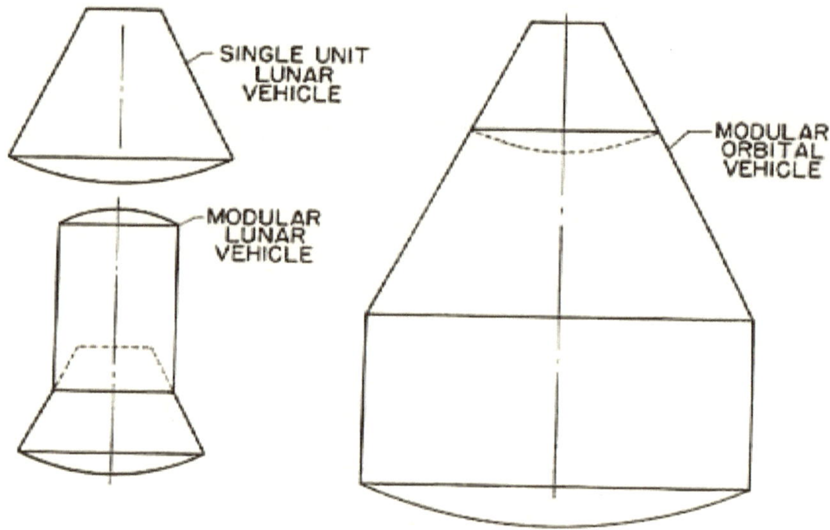

Fig. 2.7 Conceptual Vehicles

Fig 2.7 shows conceptual vehicles of components for use in both earth and lunar orbital flights. Now it is well understood that the lunar vehicle will consist of a reentry component and a constraint of minimum "mission component" to stay within the payload limitations. The "mission component" allows dropping of this portion of the vehicle prior to re-entry' thereby preventing from heating on re-entry (function?). Also this arrangement has good usable space characteristics and two compartments appear possibly desirable for any emergency condition(read Apollo 13 case below).

However studies need to be done the point where it can definitely be said that the two component vehicle is superior from all standpoints as compared to using a single, larger vehicle, such as shown in the figure above (quality function deployment?). Consequently' a detailed study needs to be made to determine the tradeoffs between using a one or two component lunar vehicle. Such a study should consider the optimum distribution of systems and supplies between the two components.

On the right side of figure 3 the complete component is illustrated, which may include either of the two sizes shown on the left, with an earth orbital laboratory or space station. The large size of the space laboratory is allowed because of the increased earth orbital payload capability as compared to lunar payload capability. It is suggested that the first of these orbiting space laboratories might be made compatible with the early version of the Saturn and, therefore, should weigh 11.34 Tonnes(Metric Tons) or less including the reentry component.

Saturn C-1. Booster

Two stage C-1 vehicles were available for research and development testing in early December 1962. Manned Vehicles qualified for use were available in 1966. Ten flights were tested.

C-1 version of Saturn did not have sufficient payload capability for the manned circumlunar mission; however, it had the capability of placing up to 11.34 Tonnes(Metric Tons)pounds of payload in earth orbit leading up to lunar missions.

Saturn C-2. The first stage of C-2 is similar to the first stage of C-1, except that its tankage may be reduced consistent with optimum propellant loading.

The second stage of C-2 is a new stage. It used four 90 Ton thrust hydrogen oxygen engines. Third stage of C-2 is the same as that used as the second stage of C-1.

C-2 vehicles were available for research and development flight tests in 1965 and for manned tests in 1967. Four flights of this vehicle; for C-2 qualification, and testing.

The needed aerodynamic capabilities of the vehicle will be determined by studying the trade offs in performance requirements of aerodynamic L/D, guidance capabilities, and the usefulness of the maneuvering rockets in making final corrections just prior to reentry, weight will be the primary requirement here, without any doubt.

Manned Flights are further constrained by biological needs of a human body for example in acceleration and momentum; that 20g in the perpendicular and 10g in the lateral are the accepted tolerance levels. In addition if tumbling, rotation or spinning, occurs during the mission, the rate should not exceed one revolution per second and the duration should not exceed 1 minute (performance requirements?).

The recycling of urine, however, appears to be worthy of investigation, 76 Kgs. of urine would be available for reclamation; this coupled with water from perspiration and respiration greatly reduced the onboard water needs with highly reliable purification system.

TABLE 2.1 MULTIMANNED SPACE FLIGHT PROBLEM AREAS.

MISSION	VEHICLE	OPERATION
SYSTEM CONCEPT	ESCAPE SYSTEM	BOOSTER REQUIREMENTS
NAVIGATION-GUIDANCE.	HEATING PROTECTION	ABORT SENSING
GROUND CONDITION REQUIREMENTS	RADIATION PROTECTION	LAUNCH REQUIREMENTS
REENTRY TECHNIQUES	NOISE	COMMUNICATIONS
ABORT PATHS	STRUCTURE	TRACKING
HUMAN FACTORS	AUXILIARY POWER	RECOVERY
	LANDING SYSTEM	TRAINING
	INSTRUMENTATION	

2.7 Interesting Case of Apollo 13

1970 Apr 11 Apollo13 lifted off from cape Kennedy Florida as Saturn V rockets sent astronauts James Lovell, Fred Haise and John Swigert towards the moon.The mission continued on smoothly until April 13 when Lovel famously uttered "Houston we have had a problem We had a main Bbus under volt" An oxygen tank had exploded crippling the spacecraft the astronauts moved into a lunar model and NASA began plotting a strategy to swing around the moon and reach earth safely.The Apollo reduced electricity and drinking water consumption and jerry rigged an adapter 9pipe duct) to remove co2 from the air.On April17 having moved into the command module for reentry the three astronauts splashed down safely in the pacific ocean capping of a remarkable journey Apollo 13 review board later pinned the accident on an electrically initiated fire in oxygen tank no.2 which had numerous faults that should have been spotted during testing They could not land on the moon.

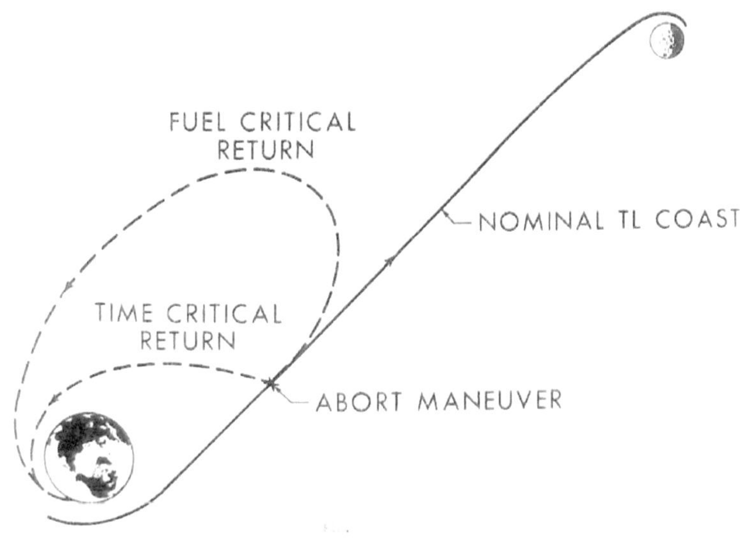

Satellite industry design up front you make careful analysis of everything you do and everything work together later on at step 8 or 9 (we are talking of our 10 step design process unit 1 p.) you end up making little bit changes you understand things are different than you modify this or that but every serious failure no one went back to the beginning and said this perhaps changed assumption on we based our design and you have to be aware of those things just to put a perspective on it. And this might make a interesting for you; if we could put everything in right perspective on a short mission!

2.8 Detailed Achievement of performance requirements

Example: Passenger lift (e.g. lift in a building)

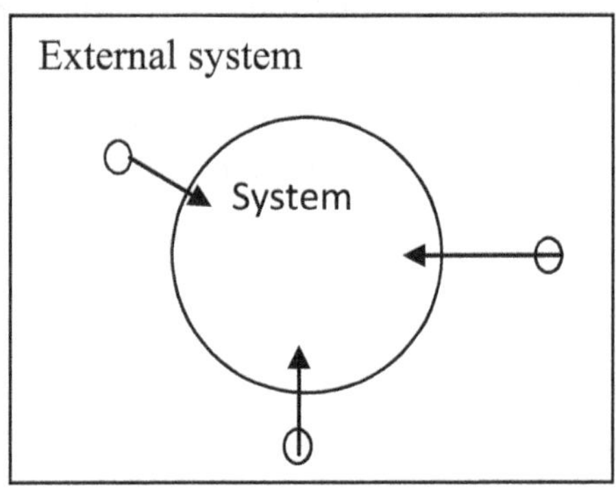

System

System task or functions

System's external systems

Systems context

External System Diagram

It is the model of interaction of the system with other system (External) in the relevant an texts this providing a definition of the system's boundary in terms of the system's input & output.

Purpose: Explicably define the system's boundary and needed interfaces.

System/ Mechanism	System Function
1. Elevators system	Provide elevator services
2. Passengers	Request and use elevator services
3. Maintenance personnel	Maintain elevator services
4. Building	Provide structural supports

NB: External System: A set of entities that interest with system....... the systems external boundaries context of a system is a set of entities that can input the system put cannot be inputted by the system.

OBJECTIVE HEIRARCHY

- Stakeholders would be willing to pay to obtain increased performance on any of these objectives.

- Developed by defining the natural Subsets of the fundamental objectives.

- Usually has two to five levels Additional information like value weight's value curves etc are added for each objective.

- Acts as a corner Stone for trade off studies.

- To be developed for each phase of the life cycle of the system.

- An important tool in decision making process.

  Operational concept Scenario-

  Scenario 1

- Passenger (including mobility, hearing, or usually challenged)
  - request up services,
  - receive feedback that their request was a accepted,
  - receive input that the elevator car is approaching,
  - and then that an entry opportunity is available,
  - enter elevator car,
  - request floor,
  - received feedback that their request as accepted,
  - receive feedback that the door is closing,
  - receive feedback about what floor the elevator stooping,
  - receive feedback that an exit opportunely is available,
  - and exit elevator with no physical in pediments.

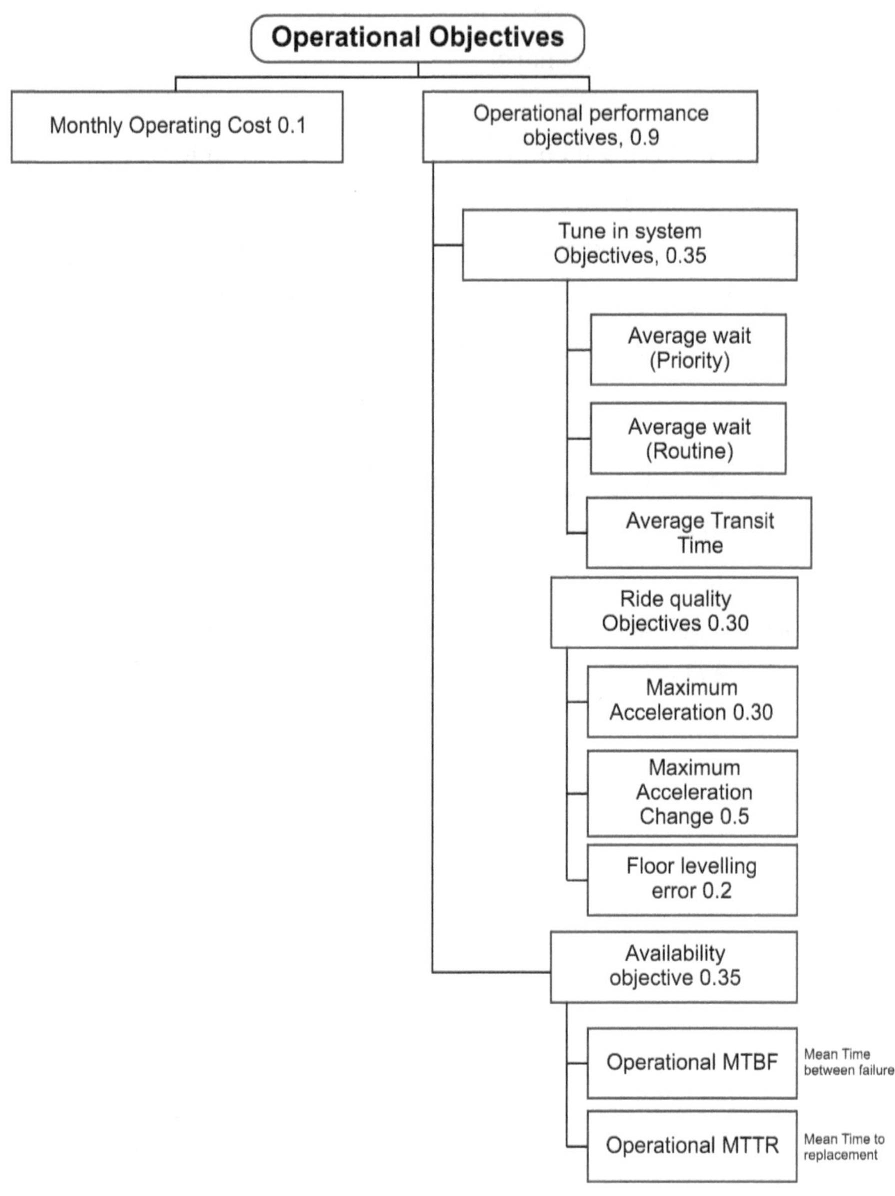

Fig.2.8 Operational objective tree for a building lift

External System Design

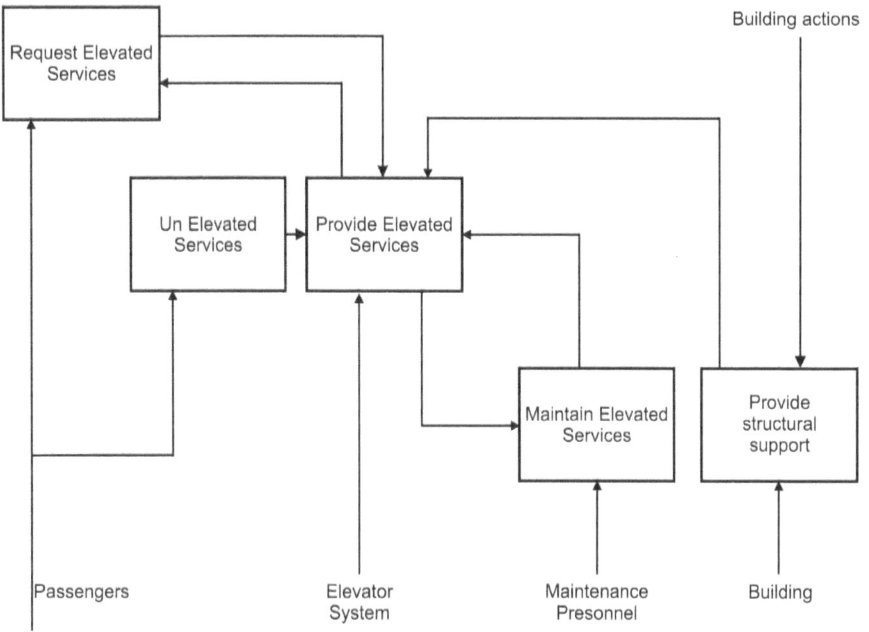

Fig.2.9 Interface with the external system.

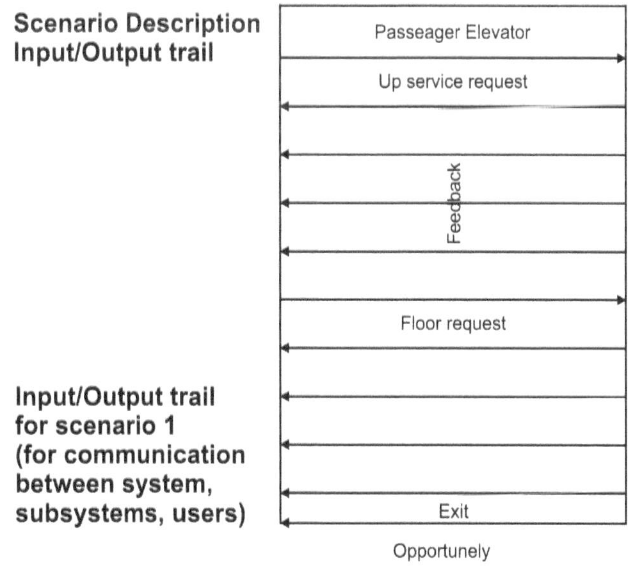

Fig.2.10 Functional decomposition diagram

Chapter 3

When you complete your study of this chapter, you will be able to:

- Understand the psychology of design (Another perspective on design process)
- Explain Modern Design team explanation of 'What is the design process?'
- Experience that our education can also be designed
- Why do we study design as a process?
- What can go wrong with a design process? Benefits of process design as a model?
- The necessity for Prescriptive Design Method or Models.
- Problem oriented approach vs. Solution oriented approach
- Comparing French Model with Pahl&Beitz Model
- Understand Various ways to visualize design
- Designers vs., Scientist Method
- Understand problem solving Methodology
- Brief on Archer's Model, VDI Model and March's Model
- Understand and differentiate between Conceptual design, Embodiment design and Detailed design.
- Understand basic characteristics of design process
- Understand factors that affect design processes and outcomes
- How is use of maths/computer involved in the design process
- System approach as analysis-synthesis (Vee approach)

After going through this critical and fundamental chapter the student must be able to built upon a solid foundation for design processes at the same time getting a threadbare information on how the modern industries develop, design and create products.

3.1 The psychology of design (Another perspective on design process)

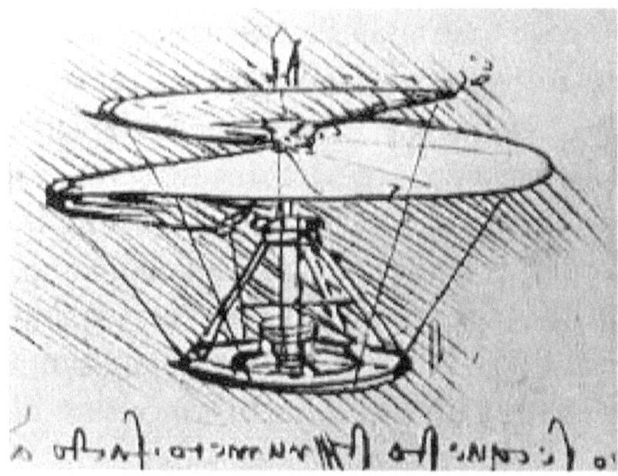

Fig.3.1 Basic helicopter drawn by Leonarda Vinci (1503AD)

Earlier engineering design was considered by many to be just a creative ad hoc process without significant scientific basis. Design was considered as an art, with successful designs created by a handful of talented individuals (example see fig.) in the same manner as great artwork was produced by talented artisans.

Fig.3.2 Tudor house in 1500 AD

However, there are now a wealth of convincing arguments that engineering design is a cognitive process; some thing that can be acquired as a skill after proper training and learning.It requires a Intelligence, very broad knowledge base and effective use of information, logics and creative thinking. Today successful designs are generated by a designing team comprising engineers, marketers, strategists, accountants, management and customers, etc., working in a structured environment and following a very systematic strategy. Utilizing information tools such as the Internet, company design documentation, brainstorming, and synergy of a design team with at times compelling and conflicting interests: information is gathered, analyzed, and synthesized with the design process yielding a final solution that meets the design criteria set by the customer.

What do we mean by a cognitive process? In the 1950s Benjamin Bloom developed a classification scheme for cognitive ability that is called Bloom's taxonomy. Figure 1.4 shows the six levels of complexity of cognitive thinking and provides an valuable insight into how the design process (following the procedure verbatim) is an effective method for designing successful products, processes, and systems. The least complex level, knowledge, is simply the ability to recall terms, concepts, principles, facts, or theories. [Who was killed in the Columbia space shuttle accident? Ms Kalpana Chawla]

The next level is Understanding, which describes the ability to understand, interpret, compare, contrast, or explain. [Explain the reason of the Columbia mishap?.Ans. insulating foam shed during launch and caused a breach in the leading edge of the left wing and thermal incompatibility of the materials caused the failure] The third level is applying, which is the ability to apply knowledge to new situations, to solve problems or tersely to generally apply the knowledge. [What would you do to prevent the repetition of Columbia like mishap ?Ans.remove material incompatibility thermally and mechanically.]

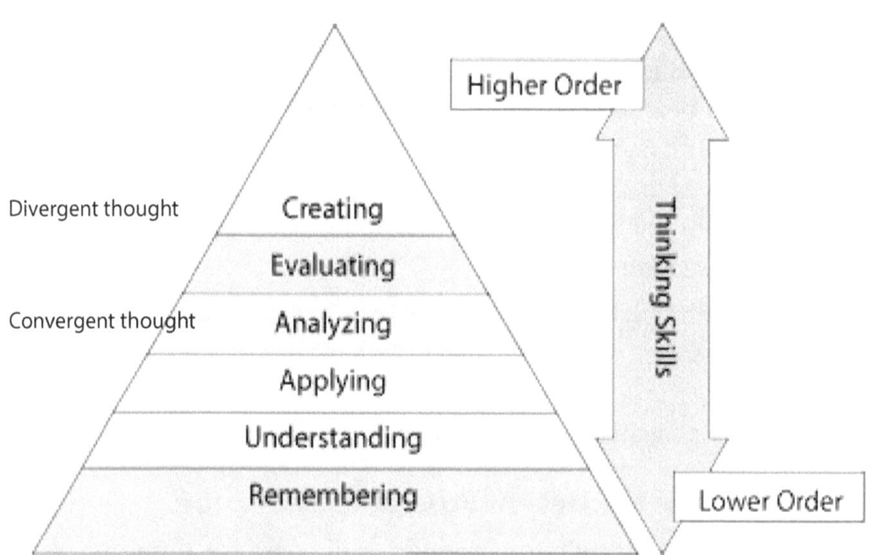

Fig.3.3 Bloom's taxonomy on learning aligns with the engineering design process

The fourth level is analyzing, which is the ability to break the already learned or acquired material into its parts, to identify the structure of its organization, to gain meaning from individual parts, relations and principles of organization. Part identification, relationships of the parts to each other and to the whole, and recognition of the principles of organization are all involved in Analyzing. The individual must understand both the content and structure of the material. Fig 3.3 shows that analysis is the highest form of convergent thought process, whereby the individual recalls and focuses on what is known and is used to solve a problem through analysis and application. [What lessons we learnt from the space program of the Columbian mishap? Improved component and product specification.]

Levels 5 and 6 on Bloom's taxonomy represent divergent thinking, in which the individual processes information and gets new insights and solutions that were not part of the original information (Thinking outside of the box).

Evaluating refers to judge quality of something based upon its adequacy, logic, value or use.

Creating (Synthesis) refers to the ability to put parts together to create something to add ideas within a solution, to come up with an action plan to formulate a new scheme.. Everyone synthesizes in a different manner. Some accomplish synthesis by quiet mental musing; others must use pencil and paper to doodle, sketch, outline ideas, and so on. [Propose an alternative to the Columbia fuel tank insulation design that would perform the required functions.]

Creation is the highest level of thinking. It is the ability to evaluate a certain idea based on a specific criteria Usually an individual is responsible for formulating the criteria to be used in the evaluation. [Assess the impact of the Columbia mishap on the U.S. space program.]

3.2 Experience the Design Process in Education

To help your understanding of the levels of cognitive thinking, review several exams you have taken in college in mathematics, chemistry, physics, and general education courses (e.g., physics, chemistry, mathematics etc). For each question, decide which level of thinking was required to obtain a successful result. You will find while moving along in your engineering curriculum that exam questions, homework problems, and projects will reflect higher and higher levels of thinking.

The design process, although structured, is an iterative process with flexibility to make necessary adjustments as the design progresses. The emphasis in this chapter is on conceptual design. At this stage of your engineering education it is important that you undergo the experience of applying the design process to a need with which you can identify based on your personal experiences. As you approach the bachelor degree you will have acquired the technical capability to conduct the necessary analyses and to make the appropriate technical decisions required for complex products, systems, and processes. Most engineering seniors will participate in a capstone design experience that will test their ability to apply knowledge toward solving a complicated design problem in their particular discipline.

You will be grilled for your intellect and creativity when you take up the major project in your coming semesters. But beware good foundations must be laid now for greater professional approach and off course the heavily credited project course.

Among the fundamental elements of the design process are the establishment of objectives and criteria, synthesis, analysis, construction, testing, and evaluation."

As an example the engineering design component of a curriculum must include most of the following features: development of student creativity, use of open-ended problems, development and use of modern design theory and methodology, formulation of design problem statements and specifications, consideration of alternative solutions, feasibility considerations, production processes, concurrent engineering design, and detailed system descriptions.

Further, it is essential to include a variety of realistic constraints such as economic factors, safety, reliability, aesthetics, ethics, and social impact.

3.3 Modern Design team explanation of 'What is the design process?

"Design is a systematic, interactive, and iterative development of a process, system or product to meet a set of specifications. It often involves working with a client to meet their perceived needs within a set of explicit and implicit constraints. It is frequently team-based and initially involves detailed needs analysis, problem definition, deployment of prior knowledge and past experience, generating and analyzing possible solutions, and exploring alternative solution paths. As the process moves forward, designers identify optimal solutions, implement selected designs, document progress, and communicate elements of the design and its process of development to multiple audiences. Often solution alternatives are derived from optimizing with respect to various resources or constraints, integrating solutions to sub-problems, and recycling components of previous designs. As they evolve, design solutions are regularly validated against performance criteria and timetables that are either agreed upon at the beginning of the process or re-negotiated at specific milestones. Throughout the entire process, participants are committed to continuous improvement in design methodology, design tools, and personal/team capabilities that are valuable to future clients."

Modern Product/System development Process

Mainly three types:

- Stage gate Process (Linear Process or waterfall Process
- -early gate ensures market for product, later gate ensures integration of Hardware and Software.
- Spiral Process (Multiple stage Gate Process)
- -Evan's spiral model for software development; also used in ship building.
- System Engineering Vee

Main characteristics of the design process are:

1. Problem partitioning and Hierarchy (Divide into smallest possible parts and components to understand it)
2. Abstraction (Tool that permits a developer to consider a component in terms of its services (behaviors)it provides without worrying about the details about its implementation. Two types of abstraction are possible functional abstraction and data abstraction.)
3. Modularity (A system is considered modular if it consists of discrete components so that each component can be implemented separately and a change to one component has a minimal impact on other components. As Jalote says "Modularity is where abstraction and partitioning come together)
4. Use of Problem space for decision making or using both Problem space and Solution space at the same time.
5. Top down and Bottom up strategies (In reality we use both these strategies and meet in the middle)
6. Distinction between iterative process and linear/waterfall process

3.4 Why do we study design as a process?

Processes of design affect the quality and reliability of our products. If we wish to better our products, we must improve our manufacturing processes besides continually redesign not just our product but also the method with which we design. That's exactly why we study the design process. To know what we do and how of our actions. To understand the method and continuously improve upon it. This is the quality of good designers.

Here, we have collected multiple descriptions of design models and processes belonging to the field of architecture to industrial design; from mechanical engineering to software development in this edition of our book.

The range of models progresses from short mnemonic devices, such as the 4Ds (define, design, develop, deploy), to very elaborate schemes, such as Archer's 229-step "systematic method for designers." Few of them look alike with the same process; but others represent very different approaches to design.

With these examples, we hope to foster debate and discussion about design and development processes; foster a breed of well skilled designers replacing a few of the also ran engineers!

How should we design? How to describe what we do? Why do we talk about it that way? How can do we do better? Why do we do it that way?

Asking these simple questions is practically relevant:

- To reduce risk (to increase the probability of success)
- To set expectations (to reduce uncertainty and fear)
- To increase iterations (to enable improvement)

Few people may benefit from deep mining into the design processes As an individual designer can be imagined as some artist working alone on music, a painting or a book; focusing on the process may not benefit him at all. But teaching new designers in working as industrial design team on a large projects requires us to reflect on each individual process (Imagine during Apollo mission 3,00,000 engineers worked at NASA). Success ultimately depends on:

- Well defined roles for each engineer and processes in advance
- Each act done by an individual is properly documented.
- Identifying and removing waste and broken processes

Ad hoc approaches to development are inefficient and cannot be repeated. There must be constant reinvention making any further improvement unthinkable. For a small scale, the costs may not be much but for large organizations (say NASA) cannot sustain prohibitive cost of incorrect design process.

Above discussion, brings to light further questions like:

How can we minimize risk at the same maximizing creativity?

The advantages of a low weightage processes over high weightage process? When will a light-weightage process is sufficient?

How to place the interaction design in respect to the larger software development process?

How software development can be integrated into the larger business formation processes?

3.5 What can go wrong with any Design Process?

- Vague or Inconsistent design process language.
- Uncertainty or lack of specificity in process.
- Branching or discussion points being unidentified
- Complexity of processes missing/not captured well.
- inappropriate details in the process.
- Tool or approaches are used that does not work
- Lack of understanding of the process by users.

3.6 Benefits of process design and modeling

- Models are faster easier and less expensive to complete.
- Models are easier to understand (as compared to other forms of documentation).
- Models provide a baseline for measurement.
- Models facilitate the use of process simulation and impact analysis of the process.
- Models use standards and use common set of techniques

Often the textbooks skip this question, what do you think of the methodology: Is the model descriptive does it describe how things are generally done here (call it descriptive model), or does it dictate to the engineer how to do the work exactly, every time (call it prescriptive model). Of Descriptive models authors does not enforce how each step should be done, and rely on smart people to make good choices. Anyway new hires need somewhere to begin, so descriptive models can be very helpful to them. A descriptive methodology might talk about what-how-build-use, or say that the organization creates a simple working system based on features and talk about the importance of prototypes. A descriptive model is supposed to model how work is actually done, so it might go so far as to say "We do X when we think it has value" or "We tend to do Y on smaller projects and Z on bigger projects." Descriptive models have drawbacks; cowboy developers can ignore them and create risk.

Prescriptive models on the contrary is categorical exactly what to do and when; Though they tend to be voluminous, require a lot of documentation, and are expensive in terms of man hours to create them. Prescriptive methodologies are like a "project insurance"; your team is ready to pay now in order to mitigate or avert a disaster later.

3.7 The necessity for Prescriptive Design Method or Models.

1. It helps us gain insight about the fundamental organization of the product systems, that is the relationship and interdependence between parts, subassembly and assembly to products (function, integration and context).

2. It helps us in understanding the relationship between a product, it's component wise make up, and it's use in larger context.

3. To use and gain competence in digital tools as a component used for prototyping.

4. To understand the underlying links between product need, material, and process.

5. Exploration of existing products and built-in technologies to develop a comprehensive understanding of the product design and integrated machinery.

6. To adopt an open ended approach to identify problems and identify opportunities for design.

"Many authors have proposed theories, models and methods in their search to explain or improve upon aspects of design practice. This field of literature, commonly known as design methodology, is primarily concerned with:...the study of how designers work and think; the establishment of appropriate structures for the design process; the development and application of new design methods, techniques and procedures; and reflection on the nature and extent of design knowledge and its application to design problems." (Cross, 1984)

"Despite the extensive research undertaken since the 1950s, there is no single model which is agreed to provide a satisfactory description of the design process" (Bahrami and Dagli, 1993). "Likewise, there is no 'silver bullet' method which can be universally applied to achieve process improvement. Instead, most methods have a well-defined and often relatively narrow focus, ranging from the generation of mechanism concepts" (e.g. Pahl and Beitz, 1996) through to the management of project risk (e.g. Baxter, 1995).

In this chapter, few of the popular approaches to the design process are presented and their practical relevance is discussed. Throughout this chapter, a classification framework is developed to support the discussion and to relate to the diverse range of forms exhibited by these models.

A process has an input and an output. Garbage in; garbage out (as is widely understood) meanwhile something may happen to the process a transformation many the transformation is reducible to a simple mathematical function. You can think this of as zip (compression) function to compress a movie. One risk in using this is that it clears a large volume but promotes an thinking of linear relationships and mechanism of cause and effect; which may not be true in certain cases. Communication

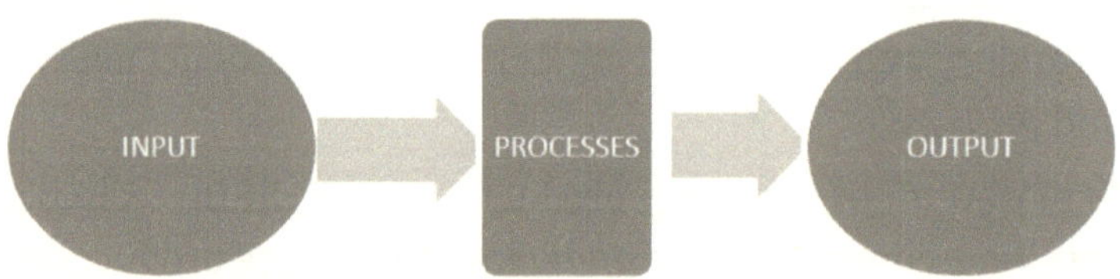

The Design Process (Reproduced from French 1999) A typical example of the more common stage-based models was proposed by French. The model, shown in Fig 3.4., is based on design practice observed in industry. It consists of four stages (French, 1999):

- Process starts with the need arising from the customer, then it is analyzed giving us a problem statement. This forms a list of requirements which the product must fulfill.

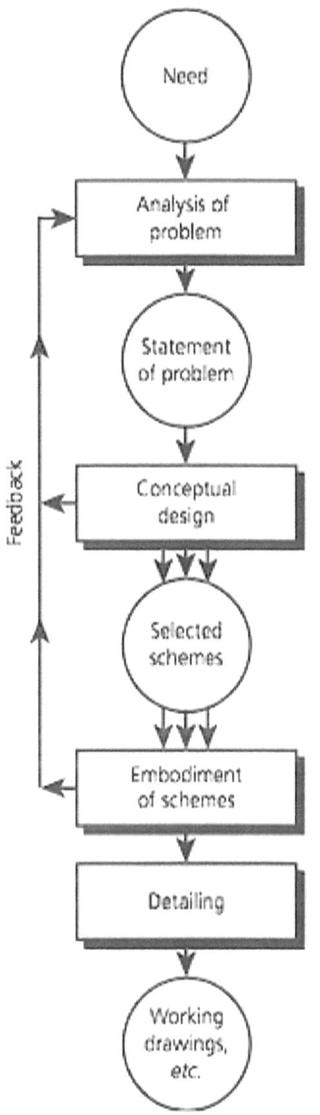

Fig.3.4 The French model of design

- Next in the conceptual design phase several concepts are developed each representing a viable set of physical idea for getting the problem solved. These schemes are then converted into a more definite representation to allow for comparison and assessment of the solutions. The resulting concepts are evaluated and one or more of them form the basis for the final solution.

- The embodiment phase consists in solidifying the chosen architecture where the abstract concept is finally transformed into a clear idea or drawing.

- The details are worked out to remove ambiguity if any from the solution, allowing the release of detailed workings, drawings for manufacturing/construction..

French acknowledges the non-linearity in his model by describing his model as hierarchical in nature; in other words, a project may encompass several stages of the model at any time according to the degree of completeness of the design.

- NB: In the diagram, the circles represent stages reached, or outputs, and the rectangles represent activities, or work in progress and squares the decision making process.

TABLE 3.1: Comparison between two design models

French (stage based)	Pahl and Beitz (phase based)
• Easy to understand and simple to use	• Complicated model, requires deeper studying
• This scheme is effective for building design.	• This scheme more effective for building design. Sharp division cannot be drawn and stages and phases do not rigidly follow one another.
• In brief the steps are state and analyze the problem, develop few conceptual designs, select the best one.	• Phase model does not show problem solving process; so in each phase designer will go through the basic design cycle more than once. All of these steps are done but with more detail.
• 'Embodiment' is too vague in the case for house-building we do it only once, so this does not seem to fit into our problem.	• Optimization takes place in this model. It is very important because there are many trade-offs available in a house design.
• Design must be as set of drawings, so that town and country planning office can pass it, and the builder can build it easily.	• Plans and drawings are more elaborated in much detail.
• The perceived 'Need' does not change with time. Design is defined here as a process to find a problem-solution pair.	• We have a fixed perceived 'Task' not changing with time. Divergence and Convergence takes place in each phase.

'Needs' often change in house building, because the loose starting specification if we were to find no solution.; this is a disadvantage here.

Model has been developed for designing new and innovative technical systems so too much attention is paid at the conceptual level and offer little procedural advice on Embodiment and detail design.

In phase-models the end of each phase can be taken as a decision point. Herein lies the importance of phase models. At the decision points you look back on the work performed, and you weigh the results obtained against the goals of the project. Phase models therefore urge a regular evaluation of the project: reject, do a step back, or continue to the following phase

According to Pahl and Beitz (1996) Phase model of the product design making the creative leap between problem definition and solution concept is the most significant problem in design or the most resistant to solution by systematic methods. It is done in the step by understanding the relationship between functional requirements and physical features of design like structural organization and mechanism amongst each part. common feature of methods attempting to support this step is an emphasis placed on the Several authors prescribe using structure requirements from which physical principles such as 'leverage' and, 'friction' are combined to solve the physical problem (develop principle solution, as in Pahl and Beitz's terminology). Functions here are an abstract formulation of the solution, not defining some possible realizations such as a 'roller bearing 'or a 'bush'. Interactions of functions in respect of transformation of energy; material and information flows provide us with an intermediate step between the problem and concept solution. Once you have ensured a function structure, a few designers advice the use of systematic consideration amongst a number of possible means for realizing each function, using morphological matrices and other combinations to ensure that most wide range of possible design concepts is covered.

Pahl and Beitz (1984) (Figure). It has the following design stages.

- Clarification of the task-To collect as much clear information about the requirements to be met in the solution and also the constraints the designer may face.

- Conceptual design- To establish the function structures in place then search for suitable solution principles and in the end combine them into variants of the concept..

- Embodiment design- The designer here determines the layout and forms and shape and develops a technical system or product cued from the concept stage in tune with the technical and cost considerations.Optimise and compare form designs. Prepare the basic part list and production documents.

- Detail design- all drawings and other production related documents are produced. Here final arrangement, form, dimensions and surface properties of all the individual parts is laid down; materials are specified; technical and cost feasibility is re-checked;

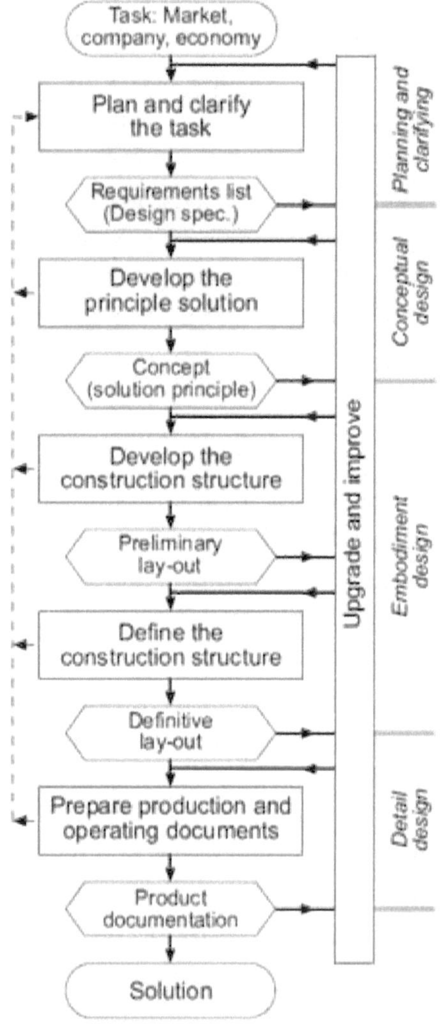

Fig.3.5 Phase model of the Product Design Process by Pahl and Beitz

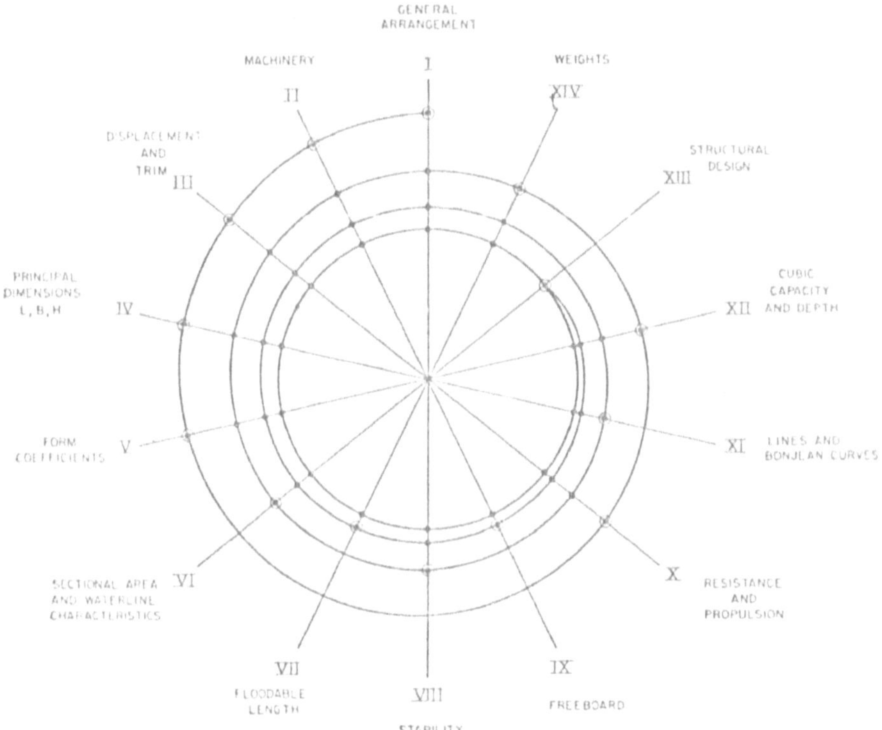

Fig.3.6 Evan's Spiral Model for Ship Designing.

As can be seen the design-focused models described above are focused mainly on the technical aspects of design problem solving, steps are either described or prescribed which are necessary to move from the problem to the solution. Some models are general in nature and have been applied over the years to many design disciplines for example Evans' spiral model is still in use for more than four decades in diverse fields ranging from software design to ship design. Few are more strongly focused on the product structure and are thus less relevant and outside of the syllabus we cover.

The intention of the methods here is to support the execution of individual design steps. They typically concentrate on the early stages of the design process ; detailed procedures for the embodiment and detail design stages are questioned by few models. Methods may be dependent on the discipline; for example morphological combination is used more in the design of non-mechanical products such as micro-processors for example. Brainstorming and requirements analysis are applicable in many situations depending on the discipline.

Another fact by Ulrich and Eppinger (2003)," who describe how bringing a screwdriver to market takes six individuals and a period of 12 months; more complex products, such as passenger aircraft, require the organization of tens of thousands of man-years' effort."

"The systematic activity necessary, from the identification of the market/ user need, to the selling of the successful product to satisfy that need— an activity that encompasses product, process, people and organization." (Pugh, 1991)

3.8 Prescriptive Models

The models are known as prescriptive in nature as they prescribe a more-or-less conventional, stepwise process of design, several attempts at building a prescriptive models of the design process are known there are books titled "101 methods of designing." Primary concern for these models is in trying to persuade and encourage designers to some standard and better ways of working. They are more systematic and their algorithmic procedure in toto is regarded as providing a particular design methodology. Prescriptive models ensure analytical work before the generation of solution concepts. This intention ensures that the design problem is t understood completely and that no over-look of important elements happens, and that the real problem is ascertained. We live in the world with many examples of good solutions to the wrong problem!

These models therefore promote a basic methodology to the design process namely analysis, synthesis and evaluation. Stages are defined as follows.

- Analysis- putting down all requirements of the problem and the break down them in to a complete set of logically related performance requirements.

- Synthesis- finding solutions to each individual performance requirements and adding the complete design elements from these with least compromise.

- Evaluation- Alternative designs are evaluated for fulfilling performance requirements for operation, manufacture and also sales before the final design are decided.

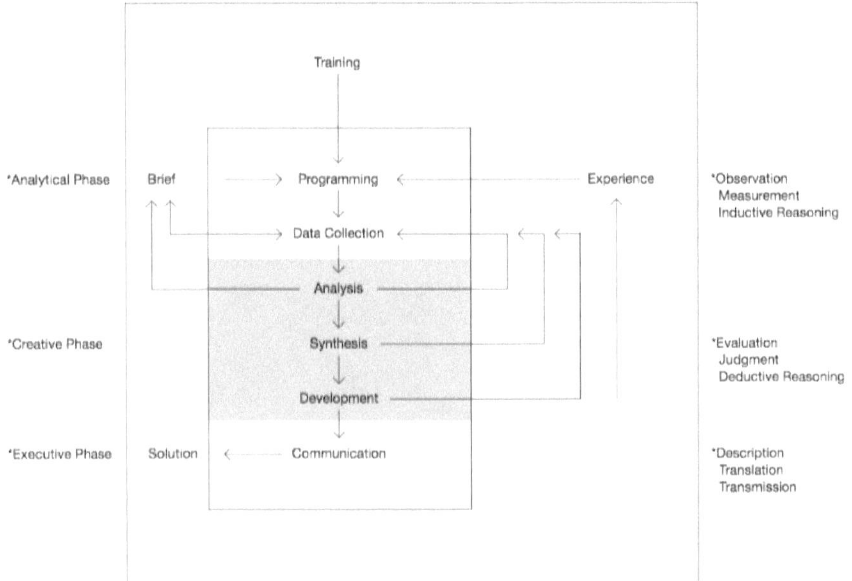

Fig.3.7 Archer's model of design process.

Though it may sound very similar to a conventional design process, but performance specifications is emphasized here logically deriving from the problem: we also generate a few alternative design concepts by building up the best sub-solutions and making a rational choice for the best amongst the alternative designs. Such apparently sensible and rational procedures are not always in practice in the conventional design methods.

In brief Archer proposed this process by dividing into three broad phases: analytical, creative and executive. Analytical phase begins with requiring objective observation and inductive reasoning, while the creative phase requires involvement, subjective judgement, and deductive reasoning; these form the backbone of the process of designing this way. Once the crucial decisions are taken, the design process continues with the execution of working drawings, schedules, etc., again in an objective and descriptive mode. The design process is thus a creative sandwich. The bread of objective and systematic analysis may be thick or thin, but the creative layer forms the center.

(1987). "Archer proposed this model as representative of an emerging "common ground" within the "science of design method" even while acknowledging continuing "differences.""

Regarding the procedure, he points out, "In practice, the stages are overlapping and often confused, with frequent returns to early stages when difficulties are encountered and obscurities found."

The practice of design is a complicated task involving contrasting skills and involves a wide field of disciplines. It requires an odd kind of mix understanding to do it successfully. Now it has become much too complicated for the designer to hold all the factors in his mind at one instance.

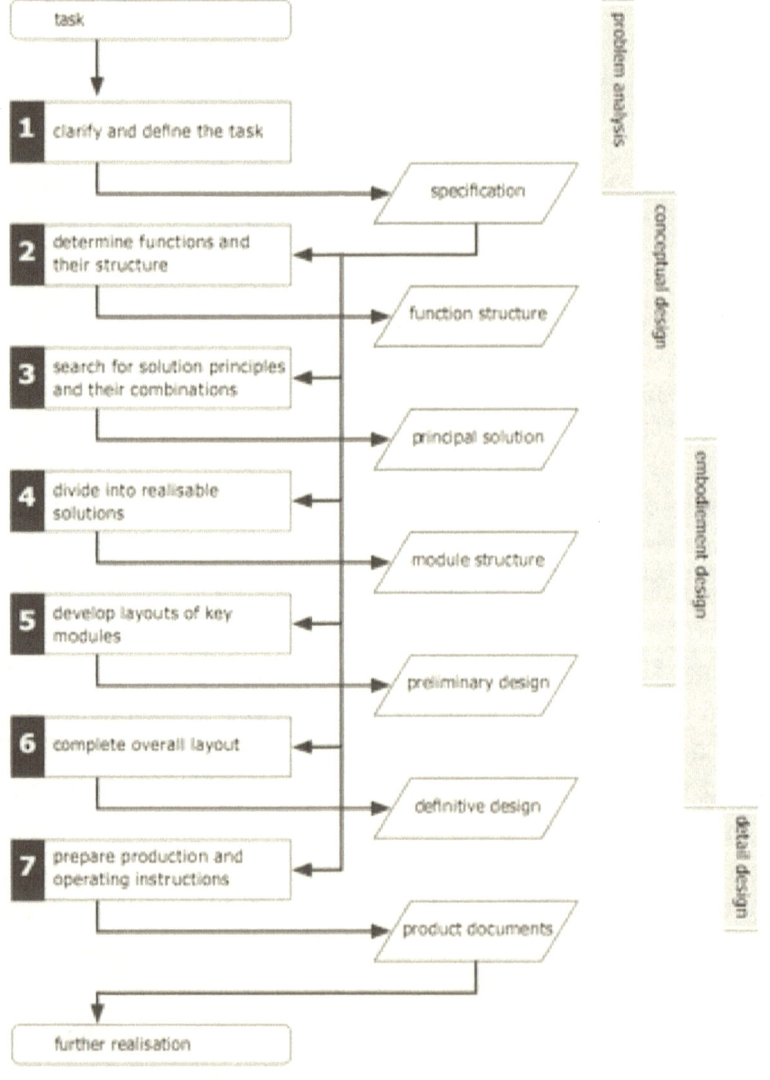

Fig.3.8 The VDI 2221 model of the design process

Germany has been on the forefront of developing several kinds of model and other aspects of rationalizing the design process. The professional engineers' society, Verein Deutscher Ingenieure (VDI), 'has produced a number of VDI Guidelines in this area, including VDI 2221: Systematic Approach to the Design of Technical Systems and Products. This Guideline suggests a systematic approach in which 'The design process, as part of product creation, is subdivided into general working stages, making the design approach transparent, rational and independent of a specific branch of industry'.

As shown in Fig.3.8 the structure of this general approach to design is build on seven stages, each stage giving a particular output.The specification which is the output from the first stage, is regarded as very important, and is reviewed constantly, updated and used as a reference in all the six stages later.

The second stage of the process consists of determining the required functions of the design, and producing a diagrammatic function structure.

A search is made for solution principles for all sub-functions in stage 3, and these are combined according to the overall function structure in the principal solution.

In stage 4, this is divided, into small modules and a module structure is made representing the breakdown of the solution into basic assemblies.

In stage 5 Key modules are developed into a set of preliminary layouts. Preliminary layouts are refined and developed in stage 6 into a definitive layout, and the final documentation is completed in stage 7.

Guideline emphasizes that analysis and evaluation of several solution variants should be done at each stage, and that there is a lot more detail in each stage than could be shown in the diagram. The following words of warning about the approach are also given by the VDI:

"It is important to note that the stages do not necessarily follow rigidly one after the other. They are often carried out iteratively, returning to preceding ones, thus achieving a step-by-step optimization. The VDI Guideline follows a general systematic procedure of first analyzing and understanding the problem as fully as possible, then breaking this into sub-problems, finding suitable sub-solutions and combining these into an overall solution. The procedure is shown diagrammatically in Figure."

This approach is based on a problem-focused, rather than a solution-focused approach. It therefore runs counter to the designer's traditional ways of thinking. so this kind of procedure has come under criticism in the design world.

A different model of the design process, which focuses the solution-focused approach of design thinking, was suggested by March in 1984 (Figure). March argued that there are two conventional forms of reasoning inductive or deductive and they are useful in evaluating and analyzing several activities of design. Now it is known that the type of activity that is most important in design is that of an act of synthesis, for which there is no established form of reasoning. March drew on the work of the philosopher Peirce to identify the missing concept as abductive reasoning. According to Peirce

*"Deduction proves that something **must be**; induction shows that something **actually is** operative; abduction suggests that something **may be**."* It is this act of synthesis, that is very much central to the design, because this kind of abductive thinking by which designs are generally produced, March calls it as productive reasoning (not abductive). Thus his model for rational design process became 'PDI model': production-deduction-induction

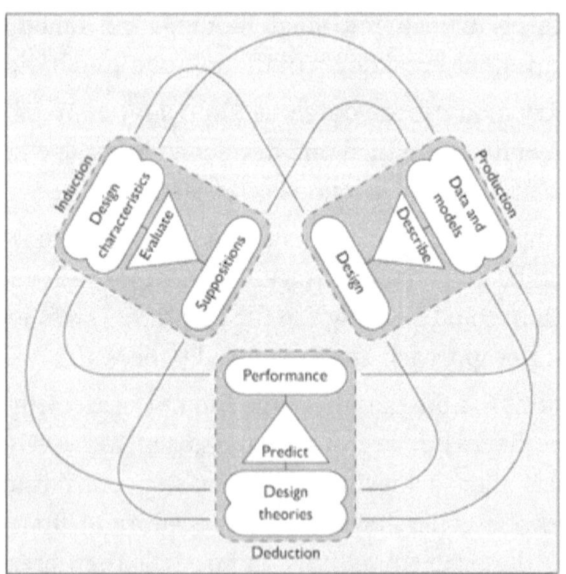

Fig.3.8 March's model of design process.

Preliminary statement of requirements, and some presuppositions about solution types in order to produce, describes a design proposal it forms the first phase of this model. (productive reasoning,) With this proposal and established theory (e.g. engineering science) now it is possible to analyze, or predict, the performance of the design (deductive reasoning). With this predicted performance characteristics it is possible inductively to evaluate further possibilities ultimately leading to changes or refinements in the design proposal.

NB. All these design processes can look quite daunting and complex but are actually more or less the same however presented from different perspectives. For more detail you can borrow some of the recommended literature from the library or search the web on the terms that have been mentioned.

3.9 Conceptual Design, Embodiment Design and Detailed design

3.9.1 Conceptual Design, Concept Development

What is a Concept?

- The concept is a plan and design which can be realized iteratively
- The concept needs to have all fundamental structural, functional and aspects of meaning* in exact order to be fulfilled in a credible and sustainable way to reach the level of independent, viable and beautiful** item, entity, product, service and/or system
- The concept has to be realized by the author, planner, designer and/or architect
- Mark the difference between example 1: idea.....|.....................concept, where a thought is on its way to direction concept) ; (example 2: idea.....................|..concept, where the sketch is a dense draft plan to be realized by fine-tuning or added add-ins and add-ons); (example 3: ....idea...........................concept, where the thought/throw is less than an idea).

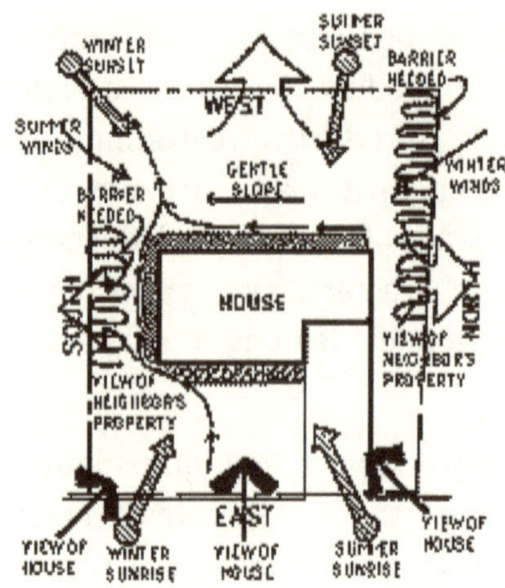

Example: Conceptual design of a House

A house and site throughout the seasons could greatly influence the overall design. The need for house not only shelter it should be energy efficient and environment friendly we are not considering structural factors and seismic concerns merely for your induction in this world of creative house planning. (example LEED see more of it on the internet. you will definitely get a divergent view to explore interesting engineering!)

So briefly factors we consider here are

- Sun and shade during different times and season

Solution: By knowing the direction of the sun at different times of the year Consume less power and water for cooling Plan open areas to allow the winter sun's rays to heat the house and outdoor living areas. Knowing that the afternoon sun is much hotter than the morning sun is also a factor to consider.

- Prevailing wind directions at different times of the day and perhaps different seasons

Solution: Prevailing winter winds can help you determine where to locate a windbreak *(which can be important if you live in the mountains).* You'll also probably not want to block summer breezes from reaching your primary outdoor living spaces in warm climates.

- Local sights and sounds
- Current soil conditions for planting appropriate trees.

Note poorly drained areas that may need underground drainage. Does water stand in low areas after a rain? Do these areas remain wet for several days? Is the soil compacted there? Does grass have trouble growing? etc., etc.

After we have completed the Research phase and we have a greater understanding in terms specifications of the product and a plan to achieve it and some needs and requirements always acting as a guide to us along the development ;it is the right time to start creating some conceptual design.

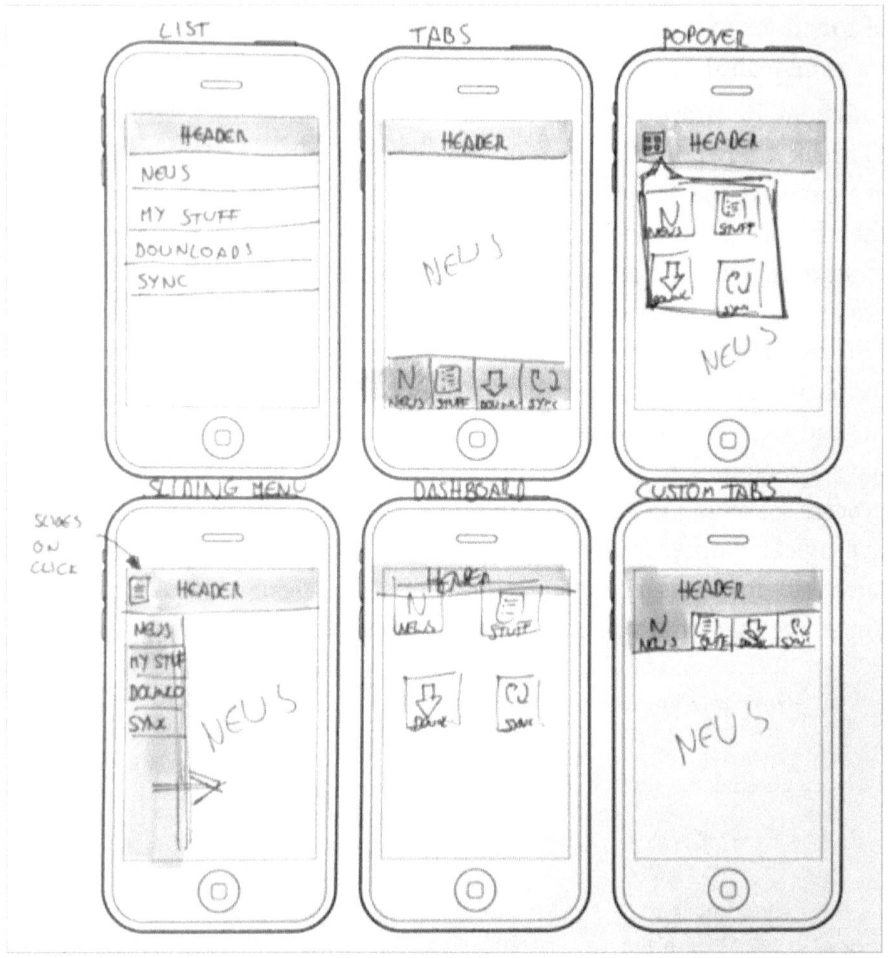

Fig.3.10 Wireframe concepts for global navigation

The main task performed in this stage is in abstracting the design in various formats. This would remove our fixation on a specific technology and conventional ideas. For example if we were using hydraulics extensively in our previous products we are not bind to use it repeatedly in our future products. Herein we emphasise what is general and essential and ignore what is particular or incidental.

Creating block diagrams of i.e. how the product layout will look like is a mandatory part of concept generation. These charts enable us to analyse the product and its inherent components and as to how they relate (or linked) to each other. The purpose of these charts should be what kind of functions and behaviours the product should inhibit and not so much focus on technical solutions.

Schematic is drawn to represent the overall functional structures and to break it down into more and more detailed sub-functions as in Fig 3.11. We begin at the System-level and break it down into sub-functions and then build up our product (Top-Down). Bottom-Up is the other option is to start off with the sub-functions that you think you will need and then create the system-level design. A guideline is to think What and not how. This requires some training since as engineers we always tend to think in terms of technical solutions that we are aware of (how things can be solved) instead of as what needs to be done (motions, cooling, heating etc). We also have a tendency to think in a timeline, in some sequential manner one process following the other or in some other way and we categorize them in a logical manner. Remember this is an iterative process and that you may have to redo things a couple of times before we (design team) are satisfied. Remember also that you can go back always and refine and change things.

Interrelationship	Elements	Structure
Functional interrelationship	Functions	Functions structure
Working interrelationship	Working principles	Working structure
Construction interrelationship	Components, Joints, Assemblies	Construction structure
System interrelationship	Artifacts, Human beings, Environment	System structure

Several systems and sub systems are identified and examined to see what kind of working principle can be used to fulfil the functions of these sub-systems. Examples of working principals for the sub-function "Energy Storage" are electrical storage, mechanical storage (flywheel, moving mass), chemical storage, hydraulic solutions (bladder, piston, Potential energy).

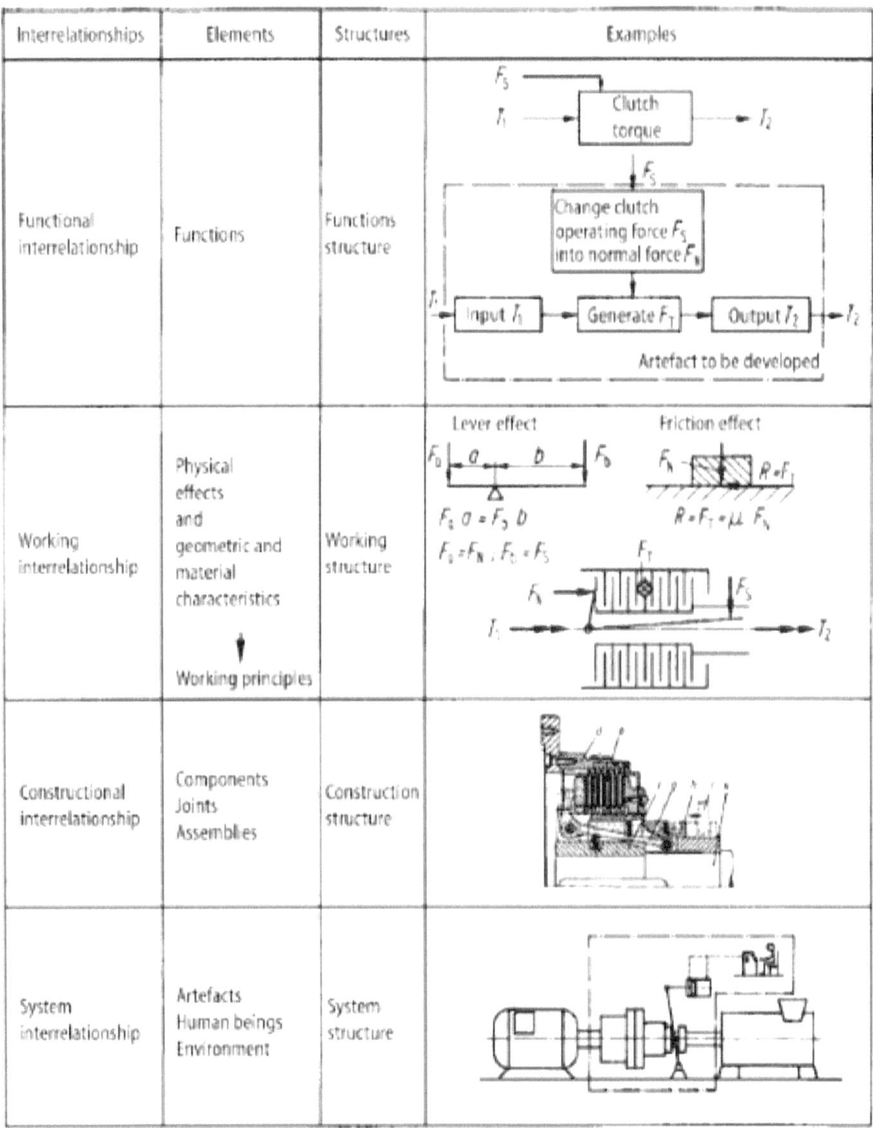

Fig.3.11 Interrelationship in electric system.

Concept Evaluation

After concept generation it is time to evaluate them. This could be handled at various levels ranging from high fidelity (very precise and with low uncertainty) to low-fidelity (gut feel, best guess etc). To compare the concepts they must be at the same level of abstraction, the options as well as the criteria to be evaluated must be in a similar language. You can't compare apples to Oranges! So we try to use measurable facts and figures (see ....as an example)

In the initial stages of concept stage of design phase uncertainty is very large and as we start refining the concepts in much greater detail you may discover that a property that you initially thought of as good changes it's behaviour during detail design. However there is always an element of uncertainty in the process just be aware of this fact and we should plan beforehand how to deal with it.

We may use different modeling techniques (physical, proof of concept or virtual models) we can use them to analyze the functionality and the criteria we have set up.

Feasibility evaluations are the best way to decide whether to go ahead or not with a technology or a concept. They are often used in the initial stages of concept evaluation. Common topics are:

"It is not feasible", *"It is conditional acceptance"*, *"It is worth considering."*

Concepts that seem not feasible to continue for development should be looked at from all other perspectives before being discarded.

There are more systematic ways of evaluating concepts via different types of Matrixes. An example of these matrixes is the decision Matrix (***Pugh's method***). This method let's you compare alternative concepts in terms of different criteria. It's a method of scoring each alternative relative to others in its ability to to meet the criteria. Comparison of the scores gives you the insight in what concept that is the best solution from a specific perspective. See Fig3.12

Software Pugh Diagram

Criteria	*	D1	D2	D3	D4	D5
Ease of Installation	9	+	-	-	/	-
Ease of Implementation	8	-	+	+	/	-
Portability	7	+	s	s	/	-
Facilitates Networking	6	+	-	s	/	s
Groups Expertise	5	+	+	+	/	-
Potential Monetary Costs	4	-	-	-	/	s
Existing code base	3	-	s	+	/	-
Long Term Maintenance	3	-	s	s	/	s
Facilitates Image Processing	2	-	+	+	/	s
Importance⁺		Scores				
	+	27	15	**18**	/	0
	-	20	19	**13**	/	32
	s	0	13	**16**	/	15

D1	Web Server	Deploying all software to AWS or RIT virtual server
D2	MATLAB Only	MATLAB Exclusivly
D3	**MATLAB GUI w/ Java**	**All components as MATLAB executables except networking components**
D4	Java GUI	Java Exclusivly
D5	C++	Complete rewrite in C++

Fig.3.12 Pugh diagram for software development (design)

Most wide used method to evaluate products and concepts is called the **Quality Function Deployment** (QFD). Very powerful if used correctly is very complex to understand thoroughly. In effect it is a combination of different matrixes that let's you correlate different requirements and functions within the product and evaluate them in terms of different criteria.

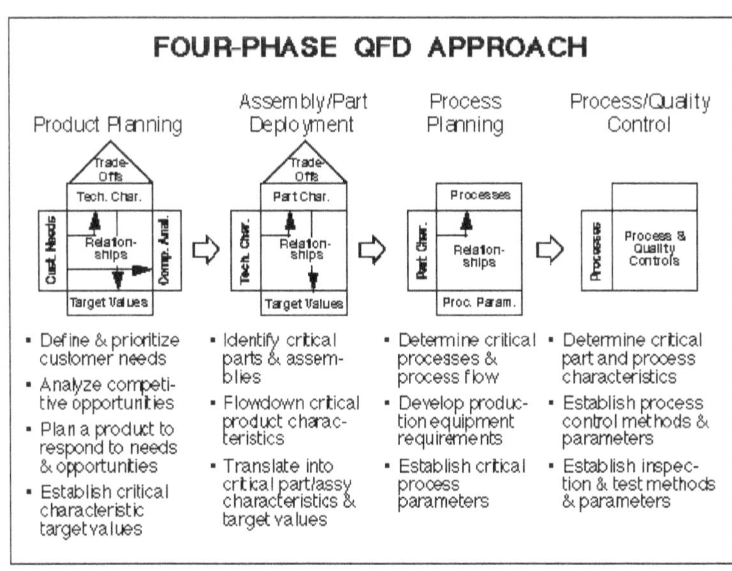

TABLE 3.2 Decision Matrix for evaluating alternate designs.

R= rating factor

R excellent 9-10, good 7-8, fair 5- 6, poor 3-4, unsatisfactory 0-2

criteria	Weight	Design 1	Design2	Design 3	Design 4
Safety	30	2	9	2	9
RX weight		60	270	60	270
Ease of use	20	8	9	6	9
RX weight		160	180	120	180
Portability	20	5	3	2	8
RX weight		100	60	40	160
Durability	10	8	8	6	8
RX weight		80	80	60	80
Standard parts	10	7	7	8	8
RX weight		70	70	80	80
Cost	10	6	5	7	8
RX matrix		60	50	70	80
Total	100	530	710	430	850

Obviously design 4 is the best.

Embodiment Design.

For elaborate answer please read sample answer on question-answer section P.180-181.

Detailed Design

This stage is involved in going from the concepts that we have created into even more detailed analysis of the parts or other components within the product or design. It aims to provide a given function with the layout, correct component shapes and material

Geometrical shape, dimensions, tolerances, surface properties and materials of the product and all is individual parts are clearly specified and laid down in the assembly drawings, detail drawings and parts lists.

- They carry instruction for production, assembly, testing, transport and operation, use, maintenance and the like, have to be worked out now. All these documents fall under the heading of the 'product documents'.

- This is the communication stage or the final stage of design

- This step is bit complex since it involves doing many things simultaneously, some steps are repeated at higher levels of information and additions and alternations in one area may well affect other components in different area.

3.10 Creative methods in modern Design

Creating concepts is a very creative process and there exist numerous methods that you can use to trigger yourself ti think in new patterns. Some of the methods used are:

- Brainstorming
- Methods 6-3-5
- Using analogies
- Using reference books and trade journals or whatever materials that you can be inspired from
- Experts – use experts from the desired field that can inspire or can come with input to your concepts. This may also involve the desired user group.

The concepts that you create may be sketches, physical prototypes of parts of the product or system or other types that explains how to solve a problem or suit a need from the requirements list.

When a number of concepts for the total system or for smaller more specific sub-functions have been created. You can try and examine if you can combine them in some way. Perhaps you can group things into a module that reduces the number of parts in the final product.

Vee System engineering approach has been represented as a case study in automobile industry for more practical relevance or purposes.

Chapter 4

Product Design Cycle,

Total life cycle,

Identification of customer needs and market research essentials,

Technology and market assessment

An exposure to various aspects of design including visual, creative and user-centric design

Visual merchandising,

trends,

materials,

technology and techniques

Various ways to think about design like visualization, photography etc,

When you complete your study of this chapter, you will be:

- Understanding the product design cycle
- Understanding the V shaped method of SDLC (Software development life cycle)
- Enumerate some good product life cycle examples
- Understanding the TPLC Total product life cycle
- Understanding the Importance of Market research in the field of design
- Understanding the process of concept generation.
- Understanding technology and market assessment through various stages of TPLC.
- Understanding Materials and Technology Inputs.
- Understanding Visual Merchandising, Trends and Techniques.
- Understanding importance of creative and user centric design
- Understand various ways to think about design visualization, photography etc.,
- Planning for Use, Distribution and retirement of product.

4.1 Product Planning and development.

After reading this chapter the student must be able to understand the effort that goes in the (Total product life cycle TPLC) management and must be able to distinguish it from the design methods and tools to make major decision during the design and development process.Offcourse a brief on other terms including customer needs, marketing and merchandising are briefly covered.

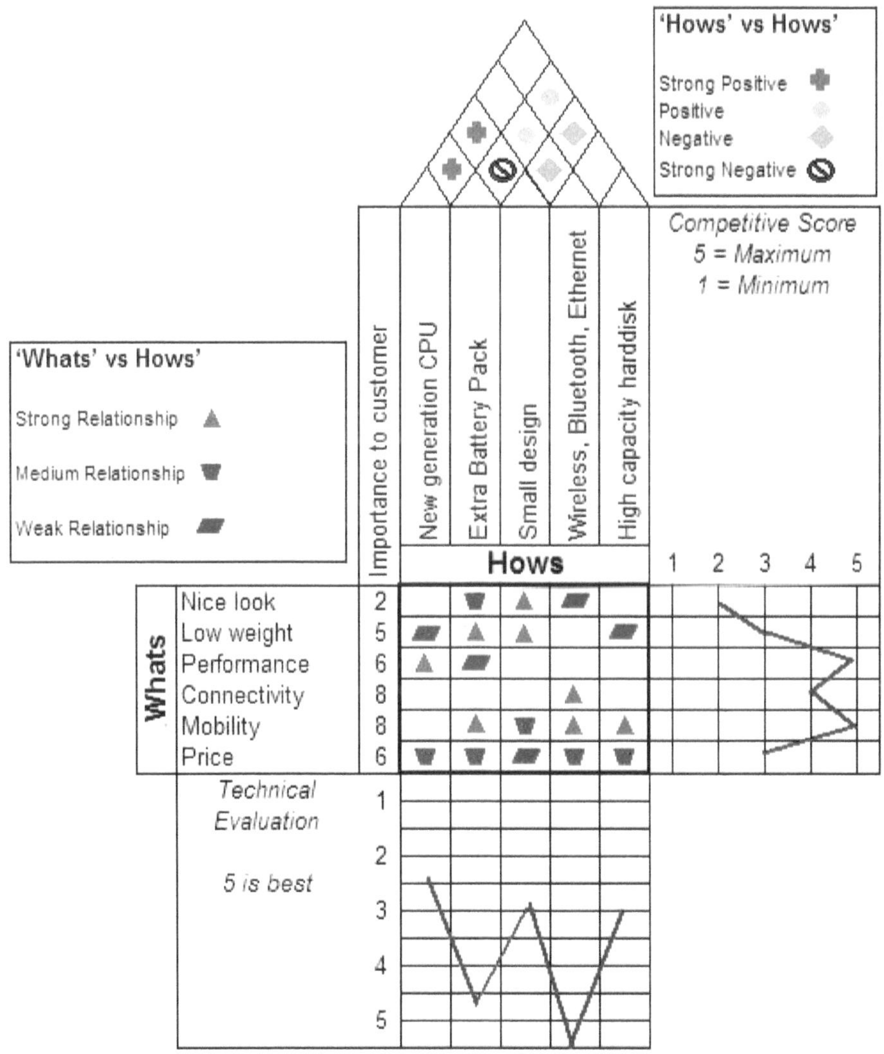

Fig.4.1 QFD (Quality function Deployment) Matrix for a mobile phone.

4.2 Importance of creative and user centric design

Often customer requirements are not well defined. The design team must determine, in consultation with the customer, the customer expectations from the solution. The customer therefore must be kept informed of the design status at all times during the process. So the process flow and information flow go hand in hand It is likely that compromises will have to be made. Both the design team and customer may trade their requirements in order to meet deadlines, cost limits, manufacturing constraints, and performance requirements. Figure 4.2 is a simple illustration of the Kano model showing the relationship between degree of achievement (horizontal axis) and customer satisfaction (vertical axis). Customer requirements are categorized in three areas:

basic, performance related, and exciting.

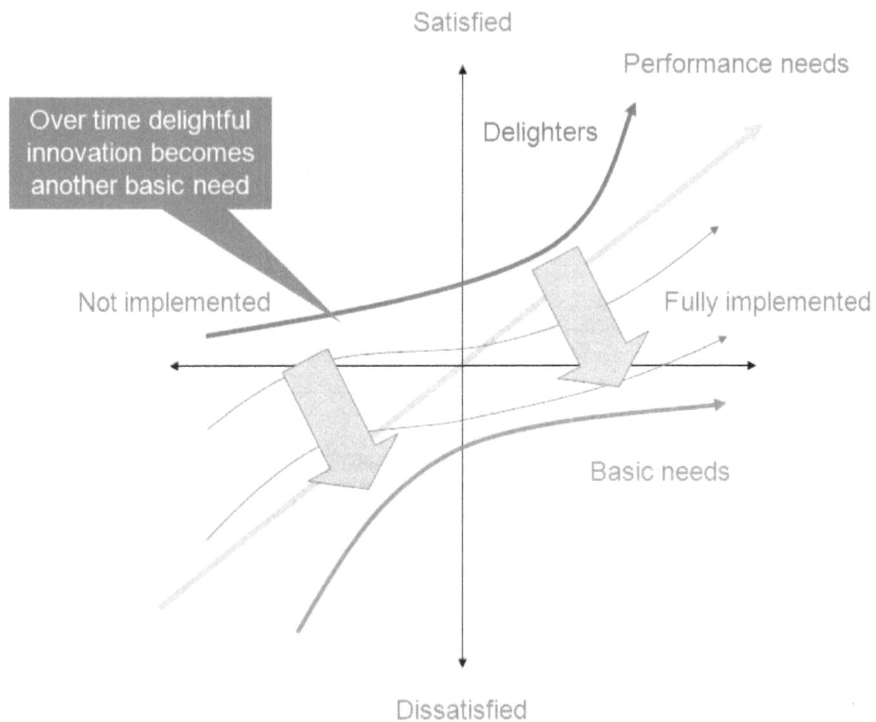

Fig.4.2 Factors in generating customer satisfaction.

Basic customer requirements are simply expected by the customer and assumed to be available. For example, if the customer desires a Mobile phone (see fig 4.1) the customer assumes that the design team and the company have proven their ability, with existing successful products, to design and manufacture mobile phones.

Performance-related customer requirements are the basis for requesting the new product. In the example of the Mobile phone nice look, low weight, performance, connectivity, mobility and price are among the many possible performance-related items that a customer may require. As time passes by and more mobile phones reach the marketplace, these requirements may become basic. Exciting customer requirements are generally suggested by the design team. The customer is unlikely to request these features because they are often outside the range of customer knowledge or vision. The exciting requirements are often a strong selling point in the design because they give the customer an unexpected bonus in the solution. Perhaps the additional capability so that mobile phone may work as a GPS device during travelling would be a unique addition to the solution; or perhaps act as a personal digital assistant (PDA) (but perhaps costlier)

Figure 4.2 indicates that the basic requirements are a must for customer satisfaction. The customer will be satisfied once a significant level of performance related requirements is met. The exciting requirements always add to customer satisfaction, so the more of these features that can be added, the greater is the satisfaction.

Many times the design team is not in touch with the customers. The customers (end users in this case) define their problem/need to the marketing or customer research firms hired by the company. This customer company interface is the most vital link for any industrial design activity. The management would then request the design team to look for solutions. The team must be careful enough not to define a solution at this step, as this would be a half baked design, it does it has not satisfied the design process.

For example, assume company management asks a team to design a cart to transport building materials from one construction site to another approximately 100 meters apart. A solution to this problem is already known: a cart. It is a matter finding how others carry their loads? what is available facility in the market? It may be to the company's benefit to find a "new system" that effectively and efficiently moves heavy loads over a short distance. Then this possibility for a rail system, belt conveyor, or any other creative solution may open up.. Usually a simple problem definition allows the most flexibility for the design team. For example, the initial problem could be defined as simply "Currently there are building material stored in building one. We need a way to get the building material from that building to the next building."

For next step (Step2) team is required to access and study all pertinent information on the problem. Internal documents of the company, widely available systems, Google searches, and other designers/engineers are all possible sources of information.

Once all team members are on board spending time on the available information, the solution constraints and criteria are identified (Step 3). A constraint is a physical or practical limitation on possible solutions; for example, the system must be operational on 240-voltAC supply. Criteria are desirable characteristics of a solution; for example, the system must be durable, maintenance free, portable, easy to operate, the system must be reliable, be, must come at an affordable cost. Constraint must be thought of as a requirement —all possible solutions must meet it (for example the system must carry a load of 100 kg per transit)—while a criterion is a relative consideration, in that one solution is better than another (for example durability is a criterion).

Next (Step 4) the design team is ready for the creative part of the process, developing alternative solutions During this step experience and knowledge, combined with group activities such as brainstorming, yield a variety of possible solutions

Each of the solutions is now analyzed thoroughly using the constraints and comparing each to the specified criteria. For few cases prototypes are built and tested to see if they meet constraints and criteria (requirements). This process heavily depends on computer modelling and analysis. Then, using a method such as a decision matrix, a solution is arrived at (Step 5).

A time schedule must be developed early in order to control the design process.

The last step of the design process consumes most time and requires resources from outside of the original design team. (Step 6) Communicating the results involves developing all the drawings, details and reports necessary for the design to be built or manufactured as well as for presentations to management and customers. Although the systematic design process appears to end at Step 6, it really remains open throughout the product life cycle. Please refer to fig 1.3 and topic 1.4 on P.6 of this book.

Technology changes, Actual test results, continuous customer feedback, and new developments in manufacturing processes, materials, changes in govt. regulations and so on may require redesign any time during the life cycle. Today many products are required to have a disposal plan prior to marketing (for example consumable spoon as a Make in India effort). In these cases the original design needs to include disposal as a constraint on the solution. Although a six (6) step design process is outlined above, other more expanded steps are in common use.

4.2.1 So do we need?

Industrial design: The professional service of creating and developing concepts and specifications that optimize the function, value, and appearance of products and systems for the mutual benefit of both user and manufacturer (from the Industrial Design Society of America's website).

Interaction design: The focus is upon how people interact with technology. The goal is to enhance people's understanding of what can be done, what is happening, and what has just occurred. Interaction designs draws upon principles of psychology, design, art, and emotion to ensure a positive, enjoyable experience.

Experience design: The practice of designing products, processes, services, events, and environments with a focus placed on the quality and enjoyment of the total experience.

Or Human Centred Design in each case?

The solution is human-centred design (HCD), an approach that puts human needs, capabilities, and behaviour first, and then de-signs to accommodate those needs, capabilities, and ways of be-having. Good design starts with an understanding of psychology and technology. Good design requires good communication, especially from machine to person, indicating what actions are possible, what is happening, and what is about to happen. Communication is especially important when things go wrong.

It is relatively easy to design things that work smoothly and harmoniously as deep consideration and study of human needs to the design process, whatever the product or service, whatever the major focus.

4.2.2 Example Functional requirements: What exactly customers want?

For complex systems involving novel technologies, people are often uncertain about what customers want, or they are unrealistic in their expectations. Many engineers avow that the most difficult task of a complex engineering project is to get the requirements right. The task depends crucially on the cooperation between systems engineers and their customers. When the customers divide into several groups, their desires often conflict say for example parents as a group may have different requirements as compared to child as a customer group. Much important for systems engineers to help their clients to clarify their objectives: What exactly do they want? Can they afford it? If not, what options do they have?

Consider an engineering project, the development of a passenger jetliner. (There are good accounts, by engineers and journalists, on the role of systems engineering in the development of the Boeing 777 jetliner in the 1990s). Systems engineers has to consider both technical and contextual conditions The major customers are the large airlines, each has its own requirements on range, speed, payload, price, fuel economy, and so on. In addition to that systems engineers also have to consult government aviation policies, airport facilities, environmental regulations, economic and demographic forecasts, and so on. To ensure that the designed airplane can be manufactured cost effectively, operated safely, traveled on comfortably, and maintained easily, systems engineers elicit opinions from manufacturers, passengers, pilots, flight attendants, and maintenance crews(Stakeholders in this case). Then they have to make tradeoff among various requirements, for instance between lower manufacturing costs and ease of maintenance for instance. Finally, having coordinated relevant requirements, they design into the airplane affordability, manufacturability, maintainability, reliability, durability, user friendless, disposability, and a host of other -abilities – dispositional properties that they anticipate will function for some time in its lifespan. They do design taking time into being, which is the gist of concurrent engineering.

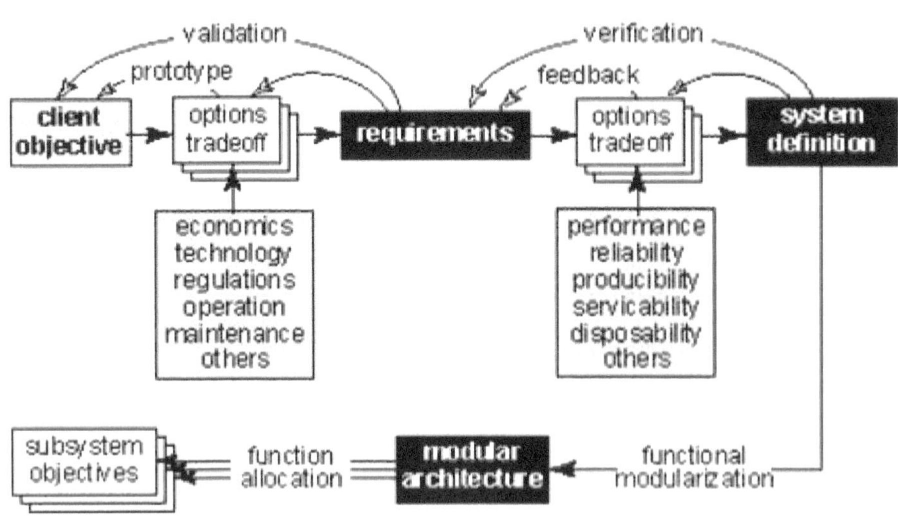

Fig.4.3 Human Centred Design (Customer focused design)

Read more on:

1. Example of a customer feedback to a website

2. IT based design or any Graphic design that can reduce the daunting and repetitive task of grocery shopping easy;in the question-answer section of this book.

4.3 SDLC (Software Development Life Cycle) Models

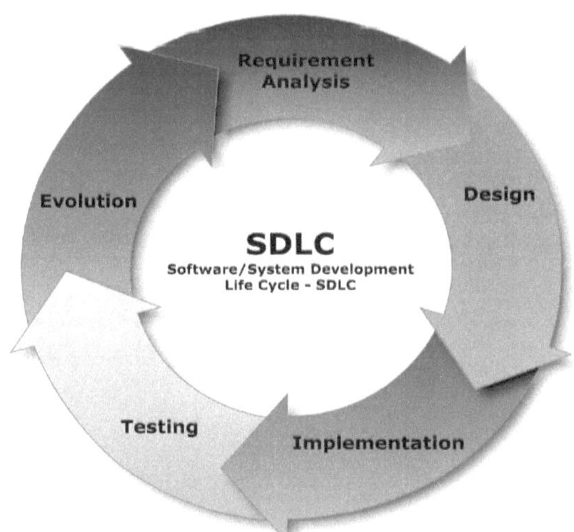

Fig.4.4 SDLC Model

Requirements-defines needed information, function, behavior, performance and interfaces

Design-data structures, software architecture, interface representations, algorithmic details

Implementation-source code, data base, user documentation, testing.

Types of SDLC Models

- Waterfall
- Iterative
- V-model
- Spiral
- Big bang
- **RAD**
- Prototyping

Each of the models for SDLC described above has its own advantages and disadvantages.

With the simplified examples of SDLC we would like to nail down the methods involved in designing product for entire life cycle as compared to product development or simple design which has some differences. Differences is in some way elaborated in Topic 4.7 and can visually be seen as difference between Fig 4.7 and Fig 4.8.

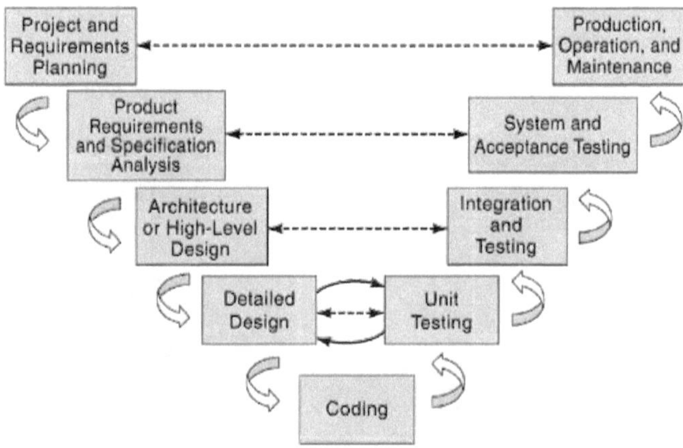

The V-Shaped Software Development Life Cycle Model

Fig.4.5 V-Shaped SDLC

4.4 Product Life Cycle Examples

A s shown in fig. traditional product life cycle curve is broken up into four key stages. Products first go through the Introduction stage, before passing into the Growth stage. Next comes Maturity until eventually the product will enter the Decline stage. These examples very well illustrate these stages for particular markets in detail.

Introduction Stage:

3D Televisions- 3D TV technology is round the corner for few decades now, but only after considerable investment from broadcasters and technology companies, 3D TVs will be available at our home.

Growth Stage:

Blue Ray Players- Advanced technology from Sony delivering the very best viewing experience, Blue Ray player is currently enjoying a steady increase in sales.

Maturity Stage:

DVD Players- Introduced in 1995 by Sony, Toshiba, Philips and Panasonic manufacturers that make DVDs, and the equipment needed to play them, have established a strong market share. However, they face challenges from other technologies like cloud space, mobiles, pen drives etc.,

Decline Stage:

Video Recorders-Beginning its journey in 1977, it is still possible to purchase a VCRs this product is definitely in the Decline Stage, as it's become easier and cheaper for consumers to switch to the other modern formats.

Another example (consumer electronics sector) also shows the emergence and growth of new technologies, and what could be the beginning of the end for those that have been around for some time.

Holographic Projection-Recently introduced into the market, holographic projection technology allows consumers to turn any flat surface into a touchscreen interface. With a huge investment in research and development, and high prices that will only appeal to early adopters, this is another good example of Introduction stage of the cycle.

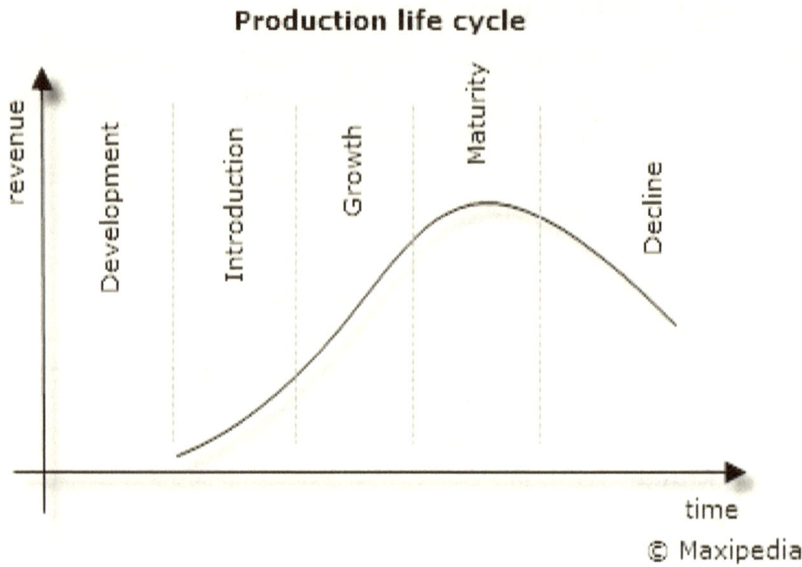

Fig.4.6 Product Life Cycle.

Tablet PCs- A variety of tablet PCs are available for consumers to choose from, as this product passes through the Growth stage of the cycle and as more competitors start to come into the market that really grew after the launch of Apple's iPad.

Laptops- Laptop computers are in use for a number of years but more advanced components, as well as additional technical up gradation that appeal to different segments of the market, will help to sustain this product as it sail through the Maturity stage.

Typewriters-Typewriters, and even electronic word processors, have very limited functionality. With consumers demanding and getting a lot more of word processing power from the electronic equipment they buy, typewriters are a product that is passing through decline stage of product life cycle.

4.5 Total Product Life Cycle Model

Since we are focused on the design concept only we are avoiding the product life cycle of development, introduction, growth, maturity and decline of any product as explained in several books as bath-tub curves etc., As the example given below will sufficiently highlight the process and principles of design only for the development stage of the or can the design model be applied to the entire life cycle we have the answer yes; as it is shown below in the case of healthcare model.

A number of safety-critical industry sectors, such as healthcare and power generation, aerospace, a formal demonstration of a product's 'fitness for use' is required before it can be released in the market. Good software development practice also places emphasis on the provision of systematic proof of fitness of function. However, this aspect of design is not explicit in many of the models presented so far. While 'evaluation' and 'iteration' are vaguely mentioned, their presence is more than a reflection of observed practice than a formal description of the steps necessary to ensure fitness for use.

(FDA, 1997). Model says very little about design, only illustrates the important and complementary roles played by verification, validation and review in the field medical device development The waterfall model has at its core a five-stage design process supported by three evaluation processes, namely:

- Verification: Establishes whether the device design described by the design output conforms to the requirements described by the design input needs.
- Validation: Establishes whether the medical device, produced in accordance with the design output, actually satisfies the users' needs.
- Review: Undertaken regularly to ensure that good practice is followed at all times.

Validation is evidently the more involved process than verification and is usually understood to be the cumulative sum of all the verification efforts.

The philosophy of validation is very much the same whether it is applied to a device or its manufacturing process. For a device, validation is ultimately achieved by showing that the final device meets the original user needs and intended uses. Process validation is achieved by showing that the process equipment meets its original needs and uses; reviewing the process equipment design and the corresponding production development simultaneously.

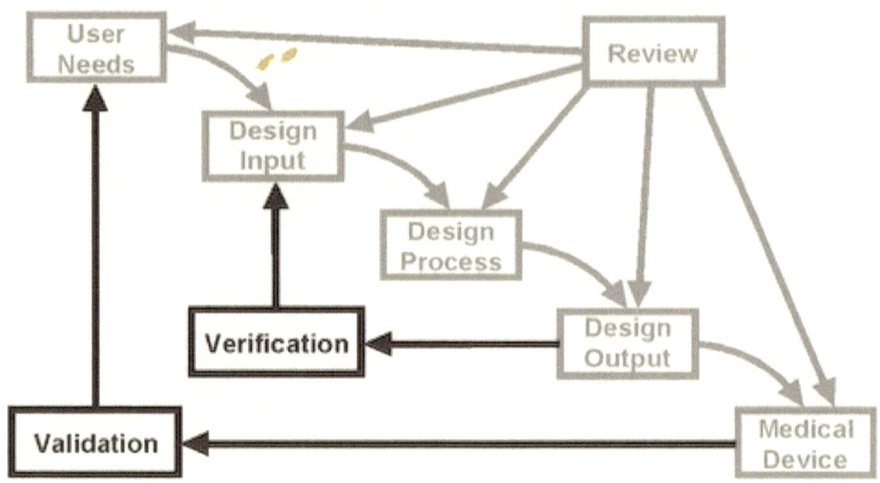

Fig.4.7 The waterfall design of Medical device design

This number of systems (·Requirements Management · Software Source Control Management · Mechanical Computer-Aided Design (MCAD) Data Management · Electrical Computer-Aided Design (ECAD) Data Management· Clinical Data Management System (CDMS)) along with known limitations of the Waterfall Model, has necessitated the development of a corrective, action-centric approach to the product design and development.

A further model of the design process as part of their strategic plan was developed by FDA, 2001. The total product life-cycle model (Fig 4.8.) is much different from the waterfall model (Fig 4.7.) and it is intended to highlight the iterative nature of device development and the connectivity between all stages of development. It refers to issues such as Design for X, verification (preclinical), validation (clinical), quality, risk management, and so on. While it does not envisage how to design a device but it does highlight many of the key issues.

TPLC model is more representative of a design process in which product development is iterative ensures both the required interactions between stakeholders and the contribution and input of every group involved in the development process (Fig 4.8 above is self-explanatory) TPLC model focuses on sharing information throughout the entire product lifecycle stages as well amongst different departments. A TPLC encourages the use of Preventative Actions over Corrective Actions, an approach that shifts focus away from rapid solution which may backfire at times. For example

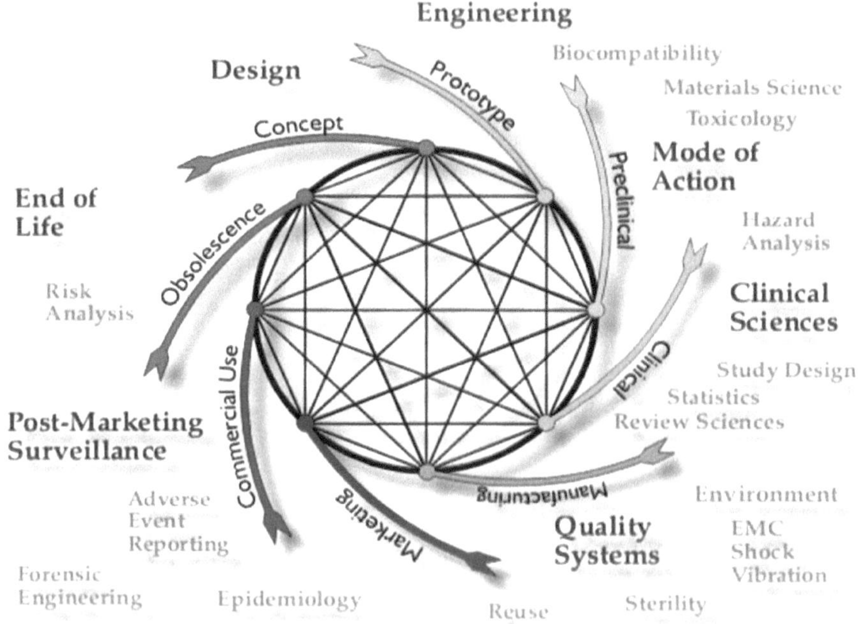

Fig.4.8 The total product life cycle model (from FDA2001)

a design engineer wants to change part for improving performance, add functionality, or to reduce cost. Upon looking for the part from the worldwide web, the engineer can view a complete history of all changes, nonconformance's, complaints associated with the part – in the true spirit of TPLC by using Windchill a renowned software for PLM (Product Life cycle Management)

Similar interesting examples can be worked out in another important fields of engineering like aerospace, power generation, defense, high-tech&electronics etc.,

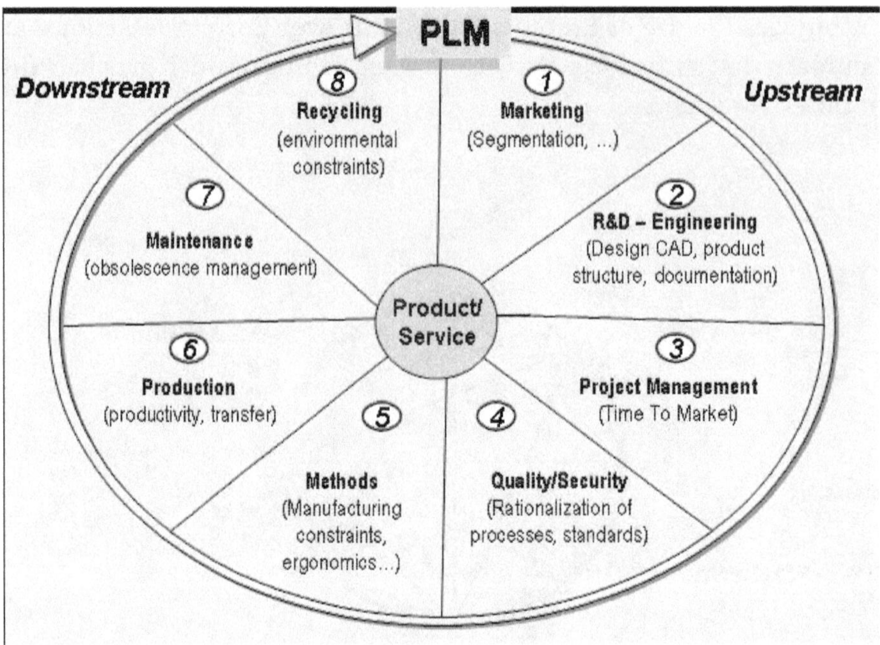

NB: PTC **Windchill** is a product lifecycle management (PLM) software product that is offered by PTC.

4.6 Product Development

In this phase you should consider **Design for X** (DfX) where X means analysing and adapting the product from different perspectives;

- Safety
- Laws and Regulations (CE-marking, ISO-9001 and ISO-14001 certificates etc.)
- Environmental impact
- Ergonomics
- Production
- Assembly
- Maintenance
- Operation
- Recycling
- Costs
- Logistics

It is important that you work in a systematic way in order not to forget something that may affect the end design negatively. Basically you look at your concepts that you still consider options for your solution and try to analyze them and adapt them according to the above topics. While you analyze your concepts you will stumble upon contradicting facts and problems, e.g. some manufacturing methods would reduce the cost of the component or product but requires a redesign of a sub-part, which will reduce the availability when assembling the product which will add to the total cost of the product. Any combination will suffice but a decision has to be made of which way to go.

However although all this analysis is done you still have to make decisions thus decision making is a large part of product development.

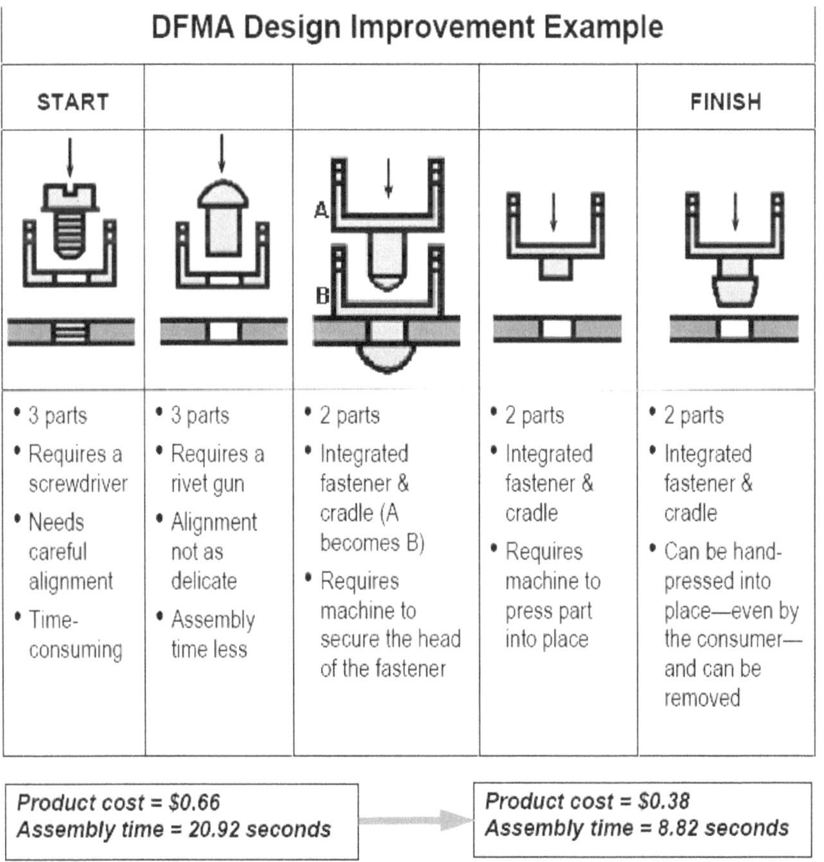

Fig.4.9 DFMA-Design for manufacturing and assembly

We work in a systematic manner so that we do not leave out something that may affect our design in a negative way. Primarily we built upon concepts that we still consider as options for our solution, analyze them and adapt them according to the design criterion (say Design For Manufacturing and Assembly (DFMA) an illustration is given for this purpose.

4.7 Design For Assembly (DFA)

A method of analyzing components/subassemblies in order to determine part relevance is called Design For Assembly. In other words DFA is the method of design of the product for ease of assembly.

DFA is involved in

- Finding theoretical minimum number of parts
- Minimizing part count/levels of assembly
- Identify the assembly process steps
- Use of Structure Charts
- Minimize levels of assembly
- Estimate the cost of assembly
- Removing non-value added process steps
- Part handling Part insertion/orientation (please refer to fig. above for clarity)

The purpose of DFA is to reduce the number of parts, optimize the assembly process and minimize assembly cost.

Basic List of DFA Methods

Minimize part numbers, part variety, and assembly surfaces; simplify assembly sequences, component handling and insertion, for faster and more reliable assembly. Aim for simplicity Every part of this document will pertain to the same idea.

- Standardize on material usage, components, and aim for as much off-the shelf component as possible to allow improved inventory management, reduced tooling, and the benefits of mass production even at low volumes. In short standardize.

- Standardize on materials, components, and subassemblies throughout product families to increase economies of scale and reduce equipment

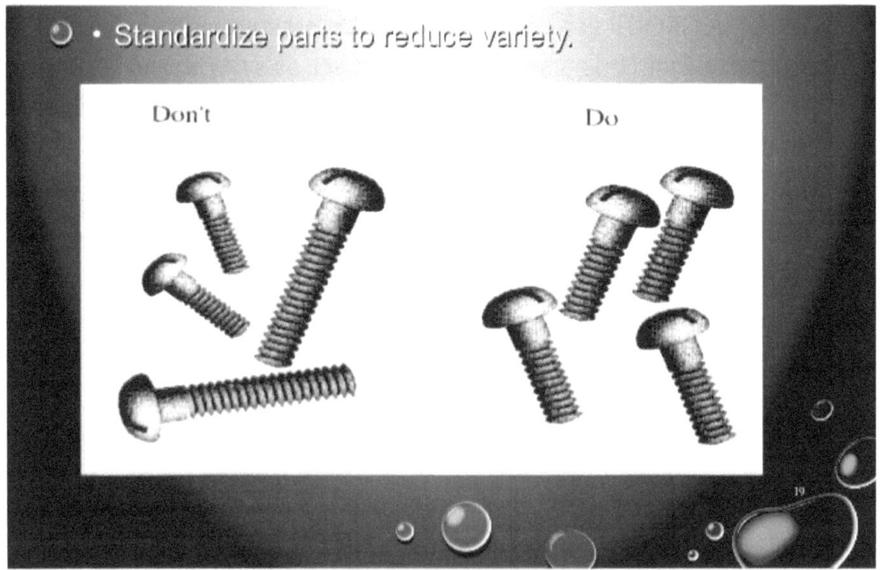

Fig.4.10 standardize parts

and tooling costs. Employ modularity to allow variety to be introduced late in the assembly sequence and simplify Just In Time production. Rationalize product design.

- Widest possible tolerances are used. Reduce the tolerance on non-critical components and thus reduce operations, and processing times.

- Choose materials to suit function and production process and to increase product reliability.

- Minimize non-value-adding operations. The minimization of handling, excessive finishing and inspection will reduce costs and lead-time.

- Design for process. (Design for Manufacturing) Take advantage of process capability to reduce unnecessary components or additional processing, for example the porous nature of sintered components is used for lubricant retention. Design in features and functions to overcome process limitations, such as features to aid mechanical feeding.

Fig 4.11 illustrates the equipment included in a typical system to meet these data transformation or number crunching requirements.

SET: A SET consists of a unit or units and the assemblies, subassemblies, and parts connected to perform a specific function. Two examples are radio receiving sets and radio transmitting sets.

GROUP:A GROUP is a collection of units, assemblies, subassemblies, and parts that (1) is a subdivision of a set or system and (2) cannot perform a complete operational function. A good example is an antenna coupler group.

UNIT:A UNIT is a combination of parts, subassemblies, and assemblies mounted together that can normally operate independently of other equipment. An example of a unit is the power supply

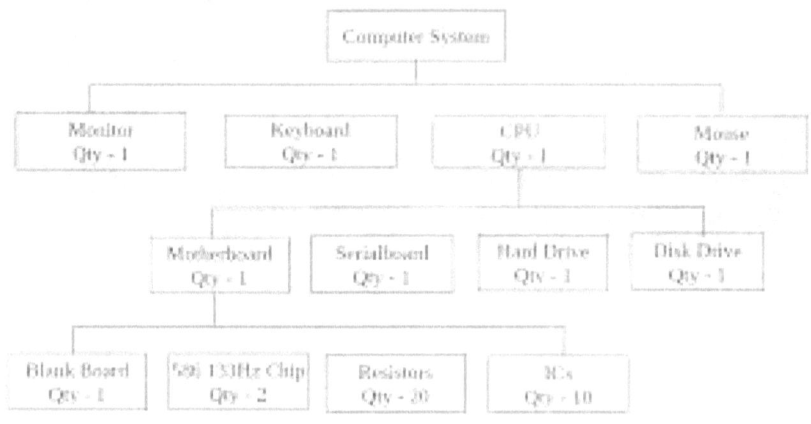

Fig.4.11 Assembly of a Computer System.

ASSEMBLY/SUBASSEMBLY: An ASSEMBLY is a combination of two or more subassemblies joined to perform a specific function. ASUBASSEMBLY consists of two or more parts that form a portion of an assembly. It can be replaced as a whole, but some of its parts can be replaced individually. The distinction between an assembly and a subassembly is not always clear. An assembly may be considered a subassembly when it is part of a larger or more complex assembly. A computer keyboard is a good example. By itself, it is an assembly. However, it is also a subassembly in a total computer system. Another example you are very familiar with is a Bicycle given below Fig 4.12.

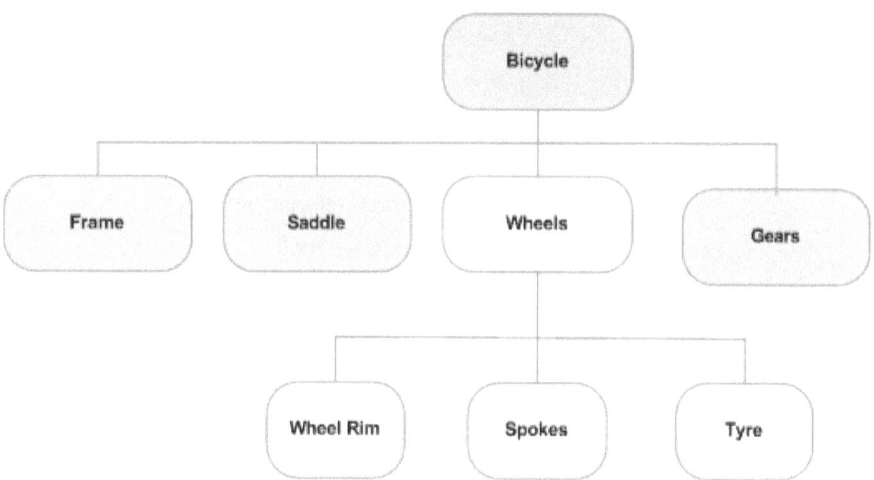

Fig.4.12 Assembly/Sub Assembly and Parts of a Bicycle.

Fig.4.13 Modular house.-DFA (Design for Assembly)?

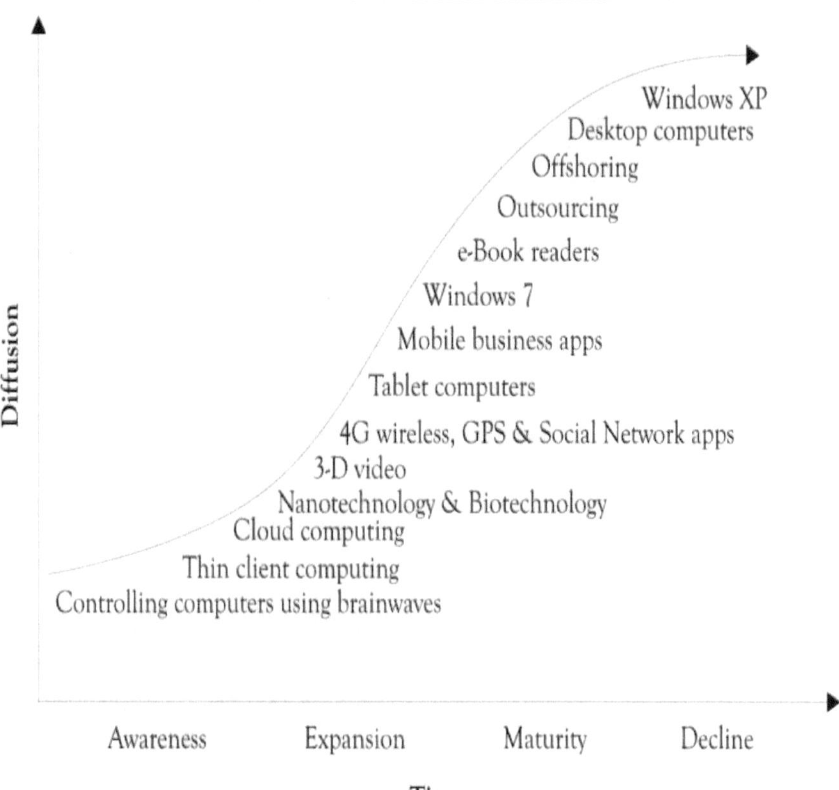

Fig.4.14 A graphical representation how product life cycle (PLC)responds to Market Diffusion.

TABLE 4.1 How Production, Marketing and other functions are defined differently from the design functions.

System design and product development Ulrich and Eppinger						
Area/ Stage	planning	Concept Development	System level design	Design detail	Tests and adjustments	Production start
Marketing	Opportunity Segment	Market needs users leader Competition	Alternatives groups	Marketing plan	Promotion Launching Field tests	Production testing with key customers
Design	Platform and architecture New technologies	Feasibility Industrial design prototypes	Alternatives Subsystems and interfaces Adjusting the industrial design	Geometry materials tolerances Industrial design	Confidence tests, life and performance certification Design changes	Evaluation of early production
Production	Specifications Strategy supply chain	Cost Feasibility	provides making Final assembly	processes Tools Quality	Production start of suppliers Setting processes Training Quality Settings	Beginning of operations

| other | technology available

financial targets

Resource Acquisition | Economic study

patents | economic feasibility of the making

Customer service | | Sales plan | |

Figure 2 A Point-Based Concurrent Engineering Process

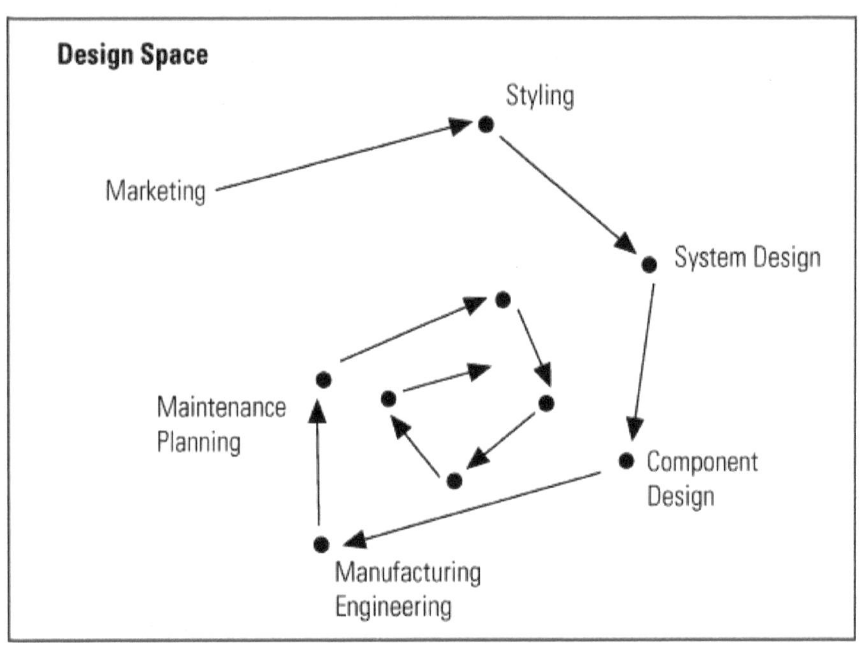

Fig.4.15 Insight as to how Manufacturing and Marketing are related to the design function.

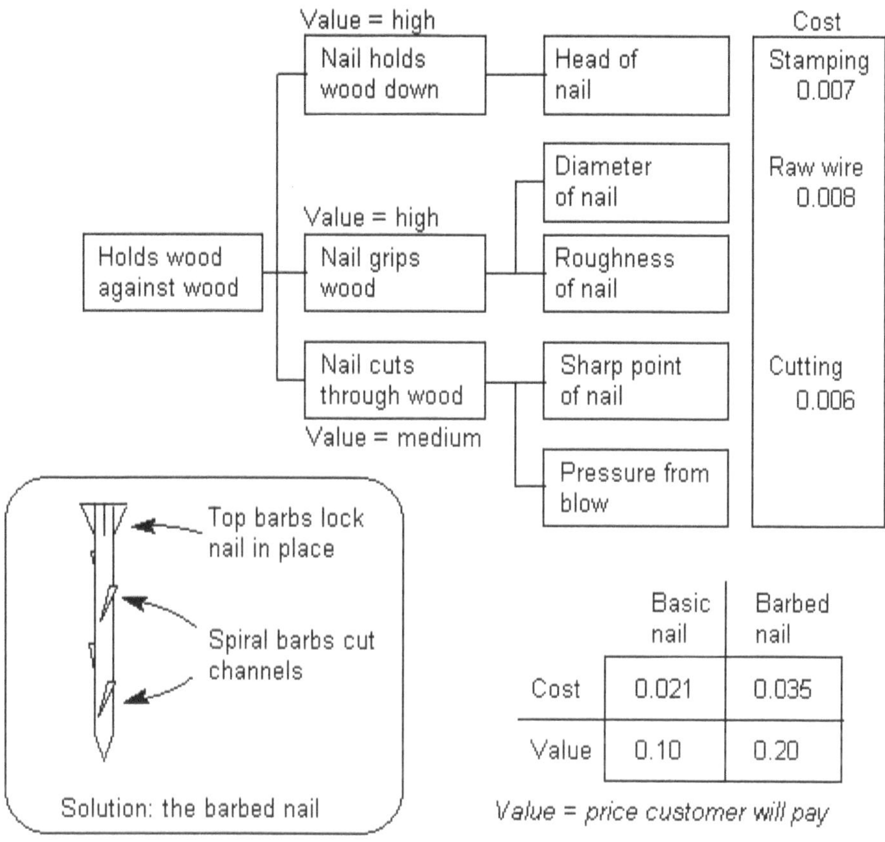

	Basic nail	Barbed nail
Cost	0.021	0.035
Value	0.10	0.20

Fig.4.16 Example Design for Value.

4.8 Materials and Technology

Recent trends in materials/mechanical field:

- Accelerate new product development
- Switch to alternate or cheaper materials
- Reduce prototyping costs
- Improve product quality and performance
- Enhance reliability
- Wide use of Composite materials, Nanotechnolgy, micromechanics of material
- Intelligent Manufacturing, laser based manufacturing
- NEMS and MEMS (Nano and Micro scale electromechanical) devices
- Shape memory Alloys and Biomechanical materials ....the list continues

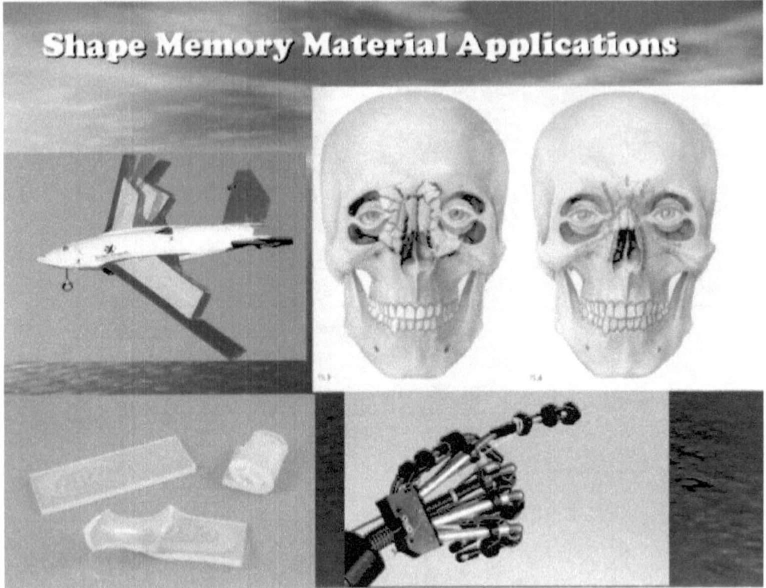

Fig.4.17 Shape Memory Material Applications.

Applications of Shape Memory Material.

Bioengineering:

Bones: Broken bones can be mended with shape memory alloys. The alloy plate has a memory transfer temperature that is close to body temperature, and is attached to both ends of the broken bone. From body heat, the plate wants to contract and retain its original shape, therefore exerting a compression force on the broken bone at the place of fracture. After the bone has healed, the plate continues exerting the compressive force, and aids in strengthening during rehabilitation. Memory metals also apply to hip replacements, considering the high level of super-elasticity. The photo above shows a hip replacement.

Reinforcement for Arteries and Veins: For clogged blood vessels, an alloy tube is crushed and inserted into the clogged veins. The memory metal has a memory transfer temperature close to body heat, so the memory metal expands to open the clogged arteries.

Dental wires: used for braces and dental arch wires, memory alloys maintain their shape since they are at a constant temperature, and because of the super elasticity of the memory metal, the wires retain their original shape after stress has been applied and removed.

Anti-scalding protection: Temperature selection and control system for baths and showers. Memory metals can be designed to restrict water flow by reacting at different temperatures, which is important to prevent scalding. Memory metals will also let the water flow resume when it has cooled down to a certain temperature.

Fire security and Protection systems: Lines that carry highly flammable and toxic fluids and gases must have a great amount of control to prevent catastrophic events. Systems can be programmed with memory metals to immediately shut down in the presence of increased heat. This can greatly decrease devastating problems in industries that involve petrochemicals, semiconductors, pharmaceuticals, and large oil and gas boilers.

Golf Clubs: a new line of golf putters and wedges has been developed using_____. Shape memory alloys are inserted into the golf clubs. These inserts are super elastic, which keep the ball on the clubface longer. As the ball comes into contact with the clubface, the insert experiences a change in metallurgical structure. The elasticity increases the spin on the ball, and gives the ball more "bite" as it hits the green.

Helicopter blades: Performance for helicopter blades depend on vibrations; with memory metals in micro processing control tabs for the trailing ends of the blades, pilots can fly with increased precision.

Eyeglass Frames: In certain commercials, eyeglass companies demonstrate eyeglass frames that can be bent back and forth, and retain their shape. These frames are made from memory metals as well, and demonstrate super-elasticity.

Tubes, Wires, and Ribbons: For many applications that deal with a heated fluid flowing through tubes, or wire and ribbon applications where it is crucial for the alloys to maintain their shape in the midst of a heated environment, memory metals are ideal.

Source:

https://depts.washington.edu/matseed/mse_resources/Webpage/Memory%20metals/applications_for_shape_memory_al.htm

TABLE 4.2 Comparison of Touch screen Material.More than often the customer demand is decisive in the use of material and technology!

5-Wire Resistive *Transmissivity/*'**Good** ____ *Clarity* ____ **75-85%**	**Capacitive (surface) Very Good** **90-98%**	**Projected Capacitive Very Good** **90-98%**	**SAW Very Good** **90-98%**	**Infrared Best** **95-100%**
***Sensor Substrate* Polyester top sheet. Glass substrate wI ITO coating**	**Glass** *wI* **ITO coating**	**Glass** *wI* **ITO coating**	**Glass** *wI* **ITO coating**	**Any Substrate**
Activate With **Any Best** *Object* **Any object** ____________	**Poor Finger or capacitive stylus**	**Good Finger, capacitive stylus, surgical glove**	**Good Finger, gloved hand, softf pliable stylus**	**Very Good Most objects**
High Sensitivity **Good** *(Light Touch)*	**Very Good**	**Very Good**	**Very Good**	**Best**
Stable *Calibration* **Very Good**	**Good**	**Very Good**	**Very Good**	**Best**
High ***Accuracy*** *e* **Very Good** ***Repeatability***	**Good**	**Best**	**Very Good**	**Very Good**
Scratch ***Resistance*** **Poor**	**Very Good**	**Best**	**Best**	**Best**
Not *Sensitive* ***to*** **Best** *Humidity* ____	**Best**	**Best**	**Very Good**	**Very Good**
Not *Sensitive* ***to*** **Best** · ***Rain/Snow*** **$**	**Average**	**Very Good**	**Average**	**Best**

Not Sensitive to Very Good **Cleaning** Chemicals	**Very Good**	**Best**	**Very Good**	**Best**
Not Sensitive **to' Best Surface Contaminants** '	**Good**	**Very Good**	**Average**	**Good**
Not Sensitive **to Best EMI/RFI**	**Average**	**Average**	**Very Good**	**Best**
Not Sensitive **to Best** Vibration ____	**Very Good**	**Very Good**	**Very Good**	**Best**
Not Sensitive **to Best Ambient** Light	**Best**	**Best**	**Best**	**Very Good**

4.9 The concept of Visual Merchandising.

Primary concept of visual merchandising is communicating with the target customer in windows and stores by attracting, connecting, and engaging with customer's through window design and product presentations to maximize sales for the store.

Visual presentation of the store plays a vital role in increasing sales of your store and can potentially convert your one time visitor as your loyal customer.It is said you never get a second chance to make a first impression.

Retailing as it stands today, attaining leadership and building an image in the customer's mind requires a lot of planning and skill. AI Rise and Jack Trout long ago talked about positioning as being a 'battle of the mind'.This is true even today: a store brand has to break through the clutter and make an impression on the customer's mind to eventually convert the image of the store that's in the customer's mind into a good image.

Image can be described as the overall look of a store and visual appeal and feelings it evokes in the eyes of the customer. The retailer wishes to develop a powerful image and provide an opportunity to embody a single message, stand out from the competition and be remembered in the mental picture.

This Image forms the foundation of all retailing efforts. Store layout, presentation, signage, events and displays can be changed reflect being new and build excitement factor from one season to the next, they must always remain true to the underlying store image.

Studies indicate that a retailer has roughly seven seconds to capture the attention of a passing customer. The following elements combine to form a distinctive image that not only reaches out and grabs the customer's attention, but also makes a positive impression in those precious few seconds.

It is important that for good sales you need to have good presentation of your store. Products placed carelessly, improper furniture, lack of cleanliness are some of the many negative factors which can send a bad image to your customer which makes it less likely for them to come back to you unless you have monopoly over your product.

Importance of Visual merchandising

Visual merchandising, also known as the 'Silent Salesman', is the science and art of suggestive selling by display and presentation.

Focal points of visual merchandising are located strategically to engage the customer in the store, and to communicate the features and benefits of the merchandise besides the in-store promotion in vogue.This is done by converting a passerby to a browser with an effective window display > a browser to a spender through the process of 'conversation' >> a spender to a big spender by increasing the 'ticket size' assisted by the process of cross-merchandising.

Visual Merchandising Enhances the Shopping Experience

Visual merchandising enhances the shopping experience by creating an image of the store in the minds of the customers besides creating a. right ambience Such an appealing ambience is created through display presentations, a combination of colors graphics, lighting, forms n fixtures.

Visual Merchandising as a Communication Tool

The right message is communicated to the customer by visual merchandising about the merchandise by projecting the latest trends, colors and fashion in apparel retailing. In retailing other merchandise, such effective visual merchandising often communicates the latest arrivals in the store. Visual merchandising, by creating basic forms, mannequins and fixtures around the merchandise, often tells the story to the customers.

Visual Merchandising Helps Customers Make Buying Decisions

Buying decisions are influenced a great deal by visual merchandising, Attimes, mere presentation creates an positive impact on the customer, who then decides to buy the product. As an example in apparel retailing, customers often ask for the whole set of outfits shown on display.

Benefits of Visual Merchandising

- Visual merchandising can induce the prospective customer to stop, look, and buy.
- Visual merchandising helps successful retailers to try an create an exclusive tangible visual identity that is synonymous with the store brand so that customers can identify with it easily.
- Visual merchandising and displays help the store cross-merchandise to create a lifestyle or to suggest complementary items, thus serving as silent salesperson.
- A store that is well-presented, with its visual merchandising and displays speaking for it, creates the right impact on the customer keeping the cash registers ringing.

Principles of Good Design for visual merchandising.

For all types of theme and type of product being sold, your store needs to be visually appealing to the masses and hence good knowledge of design principles play a crucial role.

To determine good visual displays, your graphic should embody 5 design principles:

(A) Balance

(B) Harmony

(C) Proportion

(D) Emphasis

(E) Rhythm.

Here's a brief look at each principle:

Balance – Displaying merchandise in a way that results in a pleasing distribution of visual weight within the display. This aspect is one of the most neglected aspects by people. Balancing of visual weight should be given utmost care.

Emphasis – Having a dominant point of interest in the display area, similar to a floral arrangement's focal point.

Harmony– Harmony in anything visual is desirable. your setting and arrangement of the various elements of a display, including the merchandise, props, signage and lighting, as well as colour and texture, should produce a pleasing effect.

Proportion – Keeping the ratio of one aspect of a display in relationship to the others. For example, one item shouldn't seem to be too large or small in portion to other items in the display.

Rhythm – Creating a path the eyes will follow once they've made initial contact with the item of emphasis. A display with good rhythm leads consumers' eyes throughout the entire display.

The Use of Color

Good visual merchandising also uses color to maximum effect to attract passerby's attention. As the florist is a master at using color harmonies, the same the principles for visual merchandising. For example, visual displays will use a monochromatic, analogous, Adriatic, complementary, split complementary, double complementary or tone-on-tone color scheme. As regards the display's components, the visual merchandiser shall keep in mind the merchandise and the shop's image along with the customer's like and dislikes when working with color schemes.

Principles of Good Design

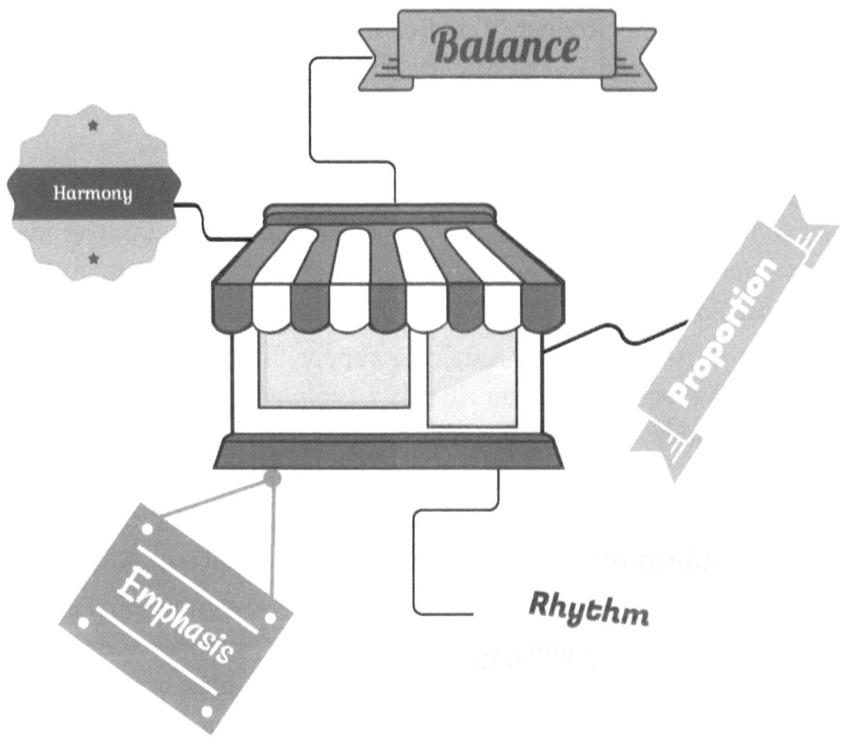

Fig.4.18 Visual Merchandising Elements.

4.10 Trends in Design.

pTC CEO Jim Heppleman set out a list of product design trends at the PTC Live Global event in Anaheim. It's useful because the interaction of these trends will impact all product designers in one way or another.

Globalization – To achieve scale economies, companies need to leverage their product development investments by delivering their products to every available market.

Localization/Mass Customization – The balancing factor to globalization is that products must be customized to meet the needs of local markets, whether it's for language, climate, culture or price. For example, John Deere ships fewer than 3 units per configuration of many tractors that sell thousands of units.

To achieve mass customization, the product has to be configurable enough to meet the needs while standard enough to be scalable.

Digitization - Companies will increasingly rely on one true digital representation of the product. That digital representation will include not just the geometry and Bill of Materials, but also the approved vendors, materials in each component, regulatory approvals, manufacturing processes and service requirements. Expect PLM to get bigger.

Regulation – Product regulation adds another layer of complexity to localization, since regulations vary by jurisdiction. For example, many jurisdictions limit the amount and source of certain materials that a product can contain.

Tracking those materials through the supply chain and then reporting on them to the regulatory authorities is a complex product management task that demands a sophisticated PLM system.

Smart products – Mechanical products aren't just mechanical anymore – they are electronic and software products as well. Even simple home appliances have lots of computing power, and this trend isn't going away.

This trend forces interdisciplinary teams and cross-functional product development. For example, Tier 1 automotive supplier Continental reportedly has more software developers than mechanical engineers.

PLM products for sophisticated manufacturers will need to build bridges between MCAD, EDA and software development platforms to coordinate the interaction among these multidisciplinary teams.

Connectivity – We've all heard of the Internet of Things. One opportunity that connected devices will create for product designers and support teams is the ability to remotely diagnose and service products in the field. That remote monitoring and serviceability in turn creates business models that don't transfer ownership, which brings us to the next big trend....

Servitization – Now there's a word that's not likely to catch on, but hey, we still say "interoperability", so who knows. In this definition, servitization represents a change in design thinking away from the physical manifestation of the product and towards conveying the benefits of the product to the end user.

Instead of buying a truck, for example, a customer can buy the miles that the truck will travel while the manufacturer retains the risks of ownership. For this to work, product developers need to integrate service lifecycle planning into their designs from the earliest concept stages.

This new world of product design will continue to favor large established multi-national players who can turn PLM technology and processes into a competitive advantage.

Some general comments on design trends in civil engineering field and software field.

1) Single-Design Model

Miller points out that traditionally, engineering documents were created then given to the contractor to re-draw with different information. Now, to save time and money, there's a shift to have it all in-house from engineering to coordination, creating a one-stop shop. Through the proper coordination, models can and should, says Miller, go through engineering right into construction. "Thirty month projects can get turned into 24-month projects," he adds.

Fig.4.18 Top trends in construction and building, from single-design models to permanent modular construction, have an impact on the future.

2) Materials

Miller, who has a degree in architectural engineering, says materials such as adapted sheet metal are popular as a guaranteed pressure class for high quality, but it's how materials are being used that makes the difference. Going to prefabrication off-site for construction fits right into the ongoing theme of improving schedules. "Instead of joining one piece of duct work you can join 20 feet of duct together," Miller says. "It's about moving more work from the field into the shop."

3) Energy Efficiency

Energy consumption is always on the mind in construction, says Miller. He notes systems that recover energy through heat wheels and occupancy sensors are becoming vital. An example of the latter are the countless interior conference rooms that can be left empty for weeks. By recognizing carbon dioxide in the room, a sensor changes the ventilation and, therefore, the energy needed.

4) Permanent Modular Construction

Snyder offers that permanent modular construction will be a huge trend in the coming years, saying the construction can easily last more than 50 years. "It looks exactly like commercial construction and can be done using many of the same things: metal studs, concrete, or even wood." Snyder says modular construction fits particularly well when you're in a time crunch, from fast food restaurants that need to go up quickly to army barracks for military deployment. "It also allows you to have an easier time doing the building as you go," he says. "Instead of building 100, 000 square feet, you can do 25, 000 and then later, add on." Snyder, who has a degree in construction science, sees it also becoming a part of high-rise construction and being particularly popular for how it fits in with LEED requirements. The key, he says, is changing people's minds about what they envision. "They see it as boxes," he says, "but it can be so many things that you want it to be."

5) Possibly... You

According to Miller, well-rounded mechanical engineers who can go beyond calculations to fitting into these coordinated engineering design models will render themselves invaluable. The only question is: Are you flexible enough for the challenge?

Flying cars, hybrid vehicles, massive jets, sleek new fighters, and Mars-bound rockets. These are the kinds of things we consider when we think of our latest heights in the endless evolution of human flight: hardware. Indeed, the old cliché about there being a million parts in an airplane is truer now than ever. But those million parts are only a fraction of the story behind what puts any vehicle in the air—and what keeps it there.

"Take a look at the cost of a Boeing 787" says Vigor Yang, chair of the School of Aerospace Engineering at Georgia Institute of Technology, Atlanta. "Fifty percent goes to hardware; fifty percent goes to navigation, guidance, and control. And of that, fifty percent goes to software."

The newest flying machines are only the most visible part of what goes on in the air. How the systems on a vehicle control that vehicle; how a vehicle talks to ground control; how a vehicle talks to other vehicles; how vehicles collect data and what they do with that data—this is the silent face of aerospace engineering. It's not tactile, it's not photogenic, and it's largely unsung. But it's where the latest advances are taking place.

1. System Software on the Rise

The code at the heart of any aircraft isn't something that can be slapped together by the latest Silicon Valley whiz kid. Unlike the programming that makes our apps and video games, airborne software is system dependent. Whoever's writing the code has got to know every aspect of the hardware. And the software must be bug free. "Otherwise everyone will be in serious trouble" says Yang. Software is handling ever-greater percentages of the jobs done on an aircraft. And, more and more, these systems are developed and put in place by companies such as Ultra-Electronics, Rockwell Collins, and Ramco Aviation. Increased communication with ground control will soon allow for more efficient landings. Currently planes approaching an airport do so in a stair-step process. This allows the control tower to maintain safety at each stage. But when the exact position of each plane is known, the approach can be continuous. The smoothness of the descent will mean every flight will be shorter by two or so minutes and save about 100 gallons of gas. That time may be minuscule for the passenger, perhaps, but worldwide, the savings are enormous.

2. Craft-to-Craft Communication

How a message gets from the cockpit to the landing gear, rudder, or anywhere else, is a relatively self-contained problem, not too different from the controls found in land-based vehicles. But how vehicles talk to each other is another issue. In a video that went viral, researchers at the University of Pennsylvania orchestrated miniature quadrotors to play the James Bond theme. The bots knew each other's location, and avoided collision, thanks to a central system that plotted their locations in space. The U.S. Air Force recently released a video showing how tiny drones will soon be able to similarly swarm together for the purposes of surveillance, targeting, and assassination. Boeing is at work creating a swarming system for larger drones. Eventually the technology will work its way into passenger planes.

3. Data Handling

Surveillance vehicles get a lot of attention for political, military, and techie reasons. But in the field of aerospace engineering their development and employment is a much smaller challenge than that of what to do with their product. How does the vast quantity of data collected from each vehicle get integrated with that from other vehicles and satellites? How does it get sifted in a way that will make it useful? How will it be streamlined and delivered to allow for effective decision-making? The answer is likely to be found with the $200 million the government recently marked for "big data" handling. Some of that will go into DARPA's XDATA program, which aims to "meet challenges presented by this volume of data" according to the Department of Defense.

4. Flying Commuters

Passenger jets and drones are not the only vehicles that will need to talk to each other in the none-too-far-off future. Though flight-minded laymen still have not seen a Jetsons-like age arrive, the personal air commute is, at least, closer than it was before. Jet pack ideas abound, (such as the Martin Jetpack and Marc Newson's "Body Jet") and flying cars are on the make (for example, Terrafugia and Moller International's Skycar). Sure, the morning commute is not likely to crowd the sky the way it does our streets anytime soon. However, if the air is thick with nine-to-fivers, there will have to be some traffic system in place. Current air-traffic control is not designed to handle localized takeoffs and landings. But, just as vehicle-to-vehicle communication is soon to keep automatic cars from colliding, aircraft-to-aircraft interaction is soon to make the man in manned aircraft a little less necessary. Congress has ordered the FAA to pave the way—legally and technically—for unmanned aircraft systems to fly in U.S. airspace by 2015. Flying commuters can piggyback on those changes.

5. Aerospace Engineering Education

Who's going to put together these systems? The kids, of course. Perhaps the biggest trend in aerospace is the growing interest among students. There are now 65 programs in the U.S., and 25 are stand alone programs. Of the 38, 000 new aerospace engineering jobs that opened up last year, 4, 000 of them were taken by students. Aerospace is the third most popular field for engineering students. A large percentage of them go into programming, "because they know their software will be implemented on real hardware" says Yang. "The aerospace profession has expanded form hardware-based science, technology, and engineering, to systems, and even systems of systems-based engineering. At a very high level that trend has become even more important" he adds.

Fig.4.19 The Martin Jetpack has a gasoline engine with two ducted fans to provide lift. Image: Martin Jetpack

Chapter 5

Evolution in Transportation and Communication Technology, Bullock Cart to Lear Jets, Personal messengers to Cell Phones, Fighter planes

Unit 5

This unit places more emphasis on the technological advancements in the field of Communication and Transportation engineering. As an engineering student is supposed to be aware in the facets of technologies which are fast changing and touches his daily life the most.

After the completion of this course the student must be able to understand the reason behind all these rationale changes lies the need or desire to solve a very imminent problem. As the country is emphasizing on both Highway and Digital way in the recent times.

5.1 Evolution in Communication Technology

- Before 3500BC – Communication relied on paintings amongst indigenous tribes.

- During 3500BC – 1099 AD there are historical proofs of accounts of trading were written on clay tablet. In India 2400: Seals made for as a signature of the writer. 2200 oldest document recorded on thick paper made of papyrus plant. 1500: Phoenician alphabet was introduced

  1400: recorded writing in China, on bones. 900: Organized postal service in China. 775: Left to right written phonetic alphabet, written from developed by Greeks 530: A collection of. 500: telesignals used including drums, shouting, trumpets, beacon fires, smoke signals, mirrors used as modes of communication in Greece. 500: Persians start a sort of pony express. 500: Chinese scholars use reeds dipped in pigment to write on bamboo. 400: Chinese use silk as well as wood, bamboo for writing material. 200: Books are written on parchment and vellum. 100: Government mail across the Roman Empire carried by couriers. 105: Paper invented by T'sai Lun. 180: An elementary zoetrope (pre film animation device) in China. 250: Usage of paper spreads to the central part of Asia. 350 Wood cover bound parchment book of Psalms in Egypt. 450: stamping Ink on seals is used on paper in China. This is a form of true copy. 600: Printing books in China. 751 other parts of world begin usage paper after a Battle 765: Printing of Picture books in Japan. 868A block-printed book in China- The Diamond Sutra 875: Chinese use of toilet paper is confirmed by travelers 950: Paper spreads more west to the Spain. 950: Folded books are used In place of rolls. 950: Women invent playing cards in a Chinese village.

- 1000-1099 1000: Writing paper made from tree bark by Mayas in Mexico, make. 1035: Japanese starts recycling used paper to make new paper.

- 1100-1199 1116: Stitched books are made by Chinese.

- 1200-1299 1200: letter system is used by European monasteries. 1200: Messenger service. started by the University of Paris 1240: A metal type is used in Korea. 1282: watermarks for the first time on paper. 1298: Use of paper money in China described by Marco Polo.

- 1300-1399 1305 – Chinese develop a wooden block movable type printing 1305: First private postal service in Europe. 1392: A type foundry is used by Koreans to produce bronze characters.

- 1400-1499 1423: Chinese method of block printing is used in Europe. 1450 – Johannes Gutenberg installs a printing press in movable type metal. 1450: Newsletters begin circulation in Europe. 1453: Printing uses metal plates 1453: 42-line Bible printed by Gutenberg 1464: A postal system is established by the king of France. 1490: Books are printed now commonly in Europe. 1495:In England the first paper mill is established.

- 1500-1599 1500: Some 10 million copies by now of 35,000 books printed. 1520 signaling each other by firing cannon and raising flags is now common by ships on the high seas. 1550: Wallpaper from China by is brought to England by traders. 1560: Precise tracing of an image is possible due to the portable camera obscura in Italy. 1560: Government controlled private postal systems come of the age in Europe. 1565: Pencil that we use today is invented.

- 1600-1699 1609: A regularly published newspaper in Germany. 1639: American colonies use their first printing press. 1650: Regular newspaper comes out in Leipzig. 1673: Between New York and Boston Mail delivery is possible. 1689: For the first time Newspapers are printed, as an unfolded "broadsides." 1696: England has 100 paper mills. 1698: Charleston, S.C.has the first Public library.

- 1700-1799 1710: First ever three-color printing developed by German r Le Blon. 1714: Henry Mill patents for a typewriter in England. 1719: Reaumur conceptualizes converting wood to make paper. 1727: Science of photochemistry begun by Schulze.1755: Between England and colonies regular mail ship is used 1770: The eraser is invented. 1774: A paper whitener invented by Swedish chemist 1775: Ben Franklin named as first Postmaster General and government. authorizes Post Office 1780: Quill feathers are replaced by Steel pen points. 1784: Vegetation instead of rags is used in French book 1785: Stagecoaches are employed to carry the mail between towns in U.S. 1790: The hydraulic press is invented in England. 1792: Mechanical semaphore signals used in France (see fig.) 1792: Britain uses Postal money orders 1792: Mail regularity is ensured by Postal Act gives throughout U.S. 1792 The first long-distance semaphore telegraph line is established by Claude Chappe 1794: Panorama opens the forerunner of movie theaters,. 1794: Signaling system establishes connection between Paris and Lille. 1798: Lithography is invented by Senefelder 1799: Robert in France invents a paper-making machine..

Fig.5.1 Semaphore signals 18 th Century France. The Semaphore arm moved to successive positions to spell out text messages in Semaphore code so the people on the next tower Kms. away could read them.

- 1800-1899 1800: Vegetable fibers used to make paper besides rags. 1801: Semaphore system is now available across the entire French coast 1804: Lithography is invented in Germany 1806: Carbon paper is produced. 1807:Improvement in Camera lucida for image tracing. 1810: Postal services consolidated within private service contracts1814 The Times in England printed by a steam-powered rotary press. 1815: Around 3, 000 post offices in the U.S. 1818: Sardinia uses stamped letter paper. 1818: Selenium is isolated ; its electric conductivity reacts to change in light1820: Arithmometer, forerunner of calculator comes into being. 1821: Sound is reproduced in England by Wheatstone 1823: Babbage builds a part of the calculating machine. 1827: Niépce develops true photograph. 1827: Wheatstone constructs a microphone. 1831 – Joseph Henry proposes and builds an electric telegraph. 1834: Babbage conceptualizes an analytical engine, forerunner of the computer. 1835 – Morse code developed by Samuel Morse 1843 – Samuel Morse installs the first long distance electric telegraph line.1837: A book on shorthand published by pitman 1838: Wheatstone's Stereoscope enables pictures in 3-D. 1838: Daguerre-Niépce method begins craze for photography 1839: Fox Talbot produces photographs in England. 1839: Jacobi invents electrotyping, the duplicating of printing plates in England. 1839: Electricity runs a printing press for the first time..1840: First postage stamps are sold in England. 1842London News appears in illustrated form. 1843: Notes explaining the computer is published by A.L.Lovelace 1844: Washington and Baltimore connected by Morse's telegraph. 1844 – Paper is produced from a wood pulp by Charles Fenerty, eliminating rag paper which was in limited supply.

- 1846: Zeiss begins manufacturing lenses in Germany. 1846: Double cylinder rotary press has a capacity of 8, 000 sheets an hour. 1849 – Associated Press arranges Nova Scotia pony express to carry latest news to readers. 1851: For the first time across the English Channel cable is laid. 1851: A flash photograph at 1/100, 000 second exposure taken by Talbot. 1853: Envelopes are made using paper folding machine. 1856:Sand boxes are replaced by blotting paper 1856:Folding paper for Newspaper, books is done by machines. 1857: Set type can be done by a machine. 1858: Eraser is placed on the pencil itself 1860: Sholes builds a functional typewriter in U.S. 1861: Telegraph technology forces Pony

Express to an abrupt end. 1861: O. W. Holmes invents stereoscope. 1862: Caselli sends a drawing over a wire for the first time. 1862: Paper money in use in the U.S.. 1864: A mail car is hooked to the railroad train for mail delivery1867: Sholes builds the functional typewriter. 1872: Simultaneous transmission is possible from both ends of a telegraphic wire. 1873: Theory of radio waves published by Maxwell. 1873: First color photographs are out 1873: Sholes' typewriters are manufactured by Remington. 1873: QWERTY pseudo-scientific keyboard is installed on the Typewriters 1876: Telephone is invented by A.G. Bell. 1878: Crookes, English chemist invents Cathode Ray Tube 1878: Hughes invents the microphone. 1880: Newspapers print photos using halftones. 1880: The electric light is invented by Edison. 1880: Leblanc conceptualizes transmitting of picture in segments. 1880: First parcel post heard off1884: People are now making long distance phone calls easily1884: Waterman's fountain pen phases out earlier versions1887: Celluloid film; is supposed to replace the glass plate photography. 1887: A multi-functional adding machine is manufactured named comptometer 1888: Picture are taken using "Kodak" box camera 1888: Demonstration of radio waves by Henry Hertz. 1888: Public telephone is now a commonplace using the coin-operation. 1888: Edison's phonograph is available for sale to the public. 1889: First automatic telephone exchange is installed by Strowger, 1892: Portable typewriters are a rage. 1892: Automatic telephone switchboard is pressded into service. 1894: Marconi invents wireless telegraphy. 1895: Paris audience sees projected movies. 1895: Friese-Greene invents phototypesetting in England 1896: New model of typewriters allows typists to see as well type at the same time called the Underwood model. 1896: New model helps sets type by machine in single characters called monotype typewriter. 1896: X-ray photography is possible..

- 1900-1999 1901: First electric typewriter is manufactured. 1901: Marconi could send a radio signal across the Atlantic. 1904: Telephone answering machine (automatic type) is invented. 1904: Radio communication is improved by diode invented by Fleming. 1904: Offset lithography becomes commercially viable. 1904: Wire can transmit a photograph in Germany. 1906 Three-element vacuum tube invented by Lee de Forest 1906: An animated cartoon film is produced for the first time.

1907: Strowger invents an automatic dial telephone switching system 1907: In Russia, Rosing develops a theory for television broadcast. 1908: Smith introduces true color motion pictures in the U, S. 1909: Disaster ship collision losses are minimized using Radio distress signal to save 1, 700 lives.1912: First mail is carried by an airplane. 1914: Reception of radio is improved by the invention of triode vacuum tube 1914: An airplane receives transmitted radio message. 1914: Leica a 35mm still camera, is launched in Germany 1914: Transcontinental telephone call is made for the first time. 1915: U.S. and Japan are connected by wireless radio service 1915: The electric loudspeaker is invented 1916: Radio is hailed as "a household utility." 1916 Optical rangefinders. are attached with camera 1916: Radios can be tuned now. 1919:Individually it is now possible to dial telephone numbers.1919: Invention of the shortwave radio. 1919:Invention of Flip-flop circuits would enable computers to count1920: The first broadcasting station come on scene. 1920: Airmail flight across the country in the U.S begins. 1920: It is now possible to record sounds electrically. 1920: KDKA in Pittsburgh broadcasts first scheduled programs. 1921: Quartz crystals helps radio signals from wandering. 1921: The word "robot" is introduced to the language. 1922: Combination of spectacles of one red and one green lens gives first 3-D movie. 1923: A picture is sent by wire for the first time after being broken into dots. 1924: 2.5 million radio sets sold in the U.S. 1925 A new standard in photography; the Leica 35 mm camera 1926: picture fax radio service is now possible across the Atlantic. 1926: Baird demonstrates the first electro-mechanical TV system. 1926: Bell Telephone Labs transmits film via television. 1927: Technicolor is invented. 1927: Negative feedback makes the possibility of hi-fi. 1928: Baird demonstrates color TV on an electro-mechanical system for the first time. 1928: Invention of the teletype machine. 1928:Three home TV setup and programming is begun. 1928: To record television Baird invents invents a video disc. 1928: Experimental TV is beamed across to the Atlantic. 1929: Phone facility for ship passengers can now phone relatives ashore. 1929: Big new: the car radio. 1929: Magnetic sound recording on plastic tape. 1929: Television studio is built in London. 1933: Armstrong invents FM, 1936: Berlin Olympics are televised using closed circuit TV. 1936: Bell Labs invents a voice recognition machine. 1936: Alan Turing's "On Computable Numbers" illustrates a general purpose computer. 1937: Digital transmission enabled by Pulse Code Modulation (PCM) 1937: Photocopier is invented by Carlsen 1937: First feature-length cartoon creation - Snow White 1938: TV in color. demonstrates live by Baird.. 1941: Moscow movie theater has the stereo effect1942: Atanasoff, Berry builds the first electronic digital computer. 1942: Kodacolor process produces the color print. 1942 – Hedy Lamarr and George Antheil invent frequency hopping spread spectrum communication technique for communication technology. 1943: Use of repeaters on phone lines obviates call noise for long distance call.1944:First digital computer Harvard's Mark I. 1945: Clarke conceptualizes geo-synchronous communication satellites. 1945: It is estimated that 14, 000 different products are made out of paper1946: Jukeboxe goes into mass production. 1946: Pennsylvania's ENIAC heralds the launch of modern electronic computer. 1947: The transistor comes on scene set to replace vacuum tubes. 1948: The LP record arrives on a vinyl disk. 1948: Shannon and Weaver of Bell Labs propound the information theory. 1948: Polaroid camera prints pictures as it takes 1949: The first real time computer is. Whirlwind at MIT 1949: Computer memory is added using. magnetic core memory 1952: Sony launches miniature version of radio transistor. 1952: EDVAC takes computer technology to new heights. 1954: Sputnik is launched by the U, S, S.R. 1954: Radio reins the world now outnumbering the newspapers. 1954: Color TV broadcasts is regular 1957: FORTRAN is now the first high-level language. 1957: First book offset printed to be entirely phototypesetter. 1958: Color can be delivered on Videotape. 1958: It is possible now to do a stereo recording 1958: Regular phone circuits can do Data transfer now1959: Invention of the the microchip. 1959 A plain paper copier is manufactured by Xerox. 1959: Artificial intelligence (AI) Bell Labs stars experimenting with it. 1959: German PAL and French SECAM formats introduced. 1960: A U.S. balloon in orbit, Echo I, reflects radio signals back down to Earth. 1962: Will operate global system. comsat set to launch, 1962: An image across the Atlantic is transmitted by Telstar satellite. 1963: Out from Holland comes the audio cassette 1963: First popular minicomputer is PDP-8 1963: Color Polaroid camera instant photograph Is available. 1964: The videotape recorder for home use from Japan. 1964: Interception of a signal off the Jupiter by Russian scientists. 1964: international satellite organization, Intelsat is finally formed. 1965: Extra services are available to customers by Electronic phone exchange 1965: Domestic TV distribution in the Soviet Union via satellites begins. 1965: Time-sharing on computer is now popular1965 - First email is sent (at MIT).1966: Multiplication of communication channels is possible by the use of Fiber optic cable 1966: Telecopier, a fax machine is launched by Xerox. 1967: Hissing noiseis eliminated by Dolby. 1967: Cordless telephones are made. 1967: Approx. 200 million telephones in the world, half of them in U.S alone 1968: Intelsat completes the desired global communications satellite loop. 1968: The RAM microchip is available now. 1969 Astronauts from the moon send live photographs 1969: Sony's U-Matic enables putting videotape on a cassette. 1969 – The first hosts of ARPANET, Internet's predecessor, are connected1971: Intel announces the microprocessor, "a computer on a chip." 1971: Wang 1200 is the first word processor in use 1972: Sony introduces 3/4 inch "U-Matic" cassette VCR1972: Erna Schneider Hoover invent a computerized switching system for telephone traffic.1972:.8-inch floppy disk removable storage medium for computers is introduced. 1972: Digital television comes out of the lab. 1974: In England, the BBC transmits the Teletext data to the TV sets. 1974Satellite begins innovative educational mission "Teacher-in-the-Sky" 1975: The microcomputer, is available in kit form in U.S. market. 1975: JVC's VHS and Sony's Betamax battle it out for customer share,. 1976: Apple I. 1976: Microprocessors control Still cameras now 1976: British TV networks also begin first teletext system. 1979: Speech recognition machine with a vocabulary of over 1, 000 words is available. 1979: Comes the digital videodisc read by laser from Holland. 1979: First cellular phone network in Japan. 1980: Sony Walkman tape player is a rage. 1980: Intelsat V relays over 2 color TV channels. and 12, 000 phone calls 1980: Atlanta gets its first fiber optics system. 1980: CNN 24-hour news channel. 1981: 450, 000 transistors can now fit on a silicon chip 1/4-inch square size. 1981: Hologram technology is improved, now in the video games. 1981: The IBM PC is launched 1981: The laptop computer is introduced. 1981: The first mouse pointing device is installed. 1982: From Japan, comes a camera with electronic picture storage, sans film. 1983: Computer chip can now hold 2.88Mbits of memory. 1983: Time names the computer as the "Man of the Year." 1984: A TV set is now introduced that can now be worn on the wrist. 1984: The 32-bit microprocessor is launched. 1985: CD-ROM can put equivalent of 270, 000 papers of text on a CD record. 1985: Cellular telephones are in the cars1990: Videodisc returns with a laser form. 1991: 3 out of 4 U.S. homes own VCRs; fastest selling domestic appliance in the entire history. 1992:Sending the first SMS (or text message)over Internet2 organization is in place.IBM ThinkPad launched. It is now lightweight compared to its predecessors.1994: U.S. government privatizes Internet management. 1995: CD-ROM disk can carry a full-length feature film for the first time. 1995: Sony demonstrates the flat TV set. 1995: Plan to put greater part of the nation on-line within 5 years announces Denmark. 1995: Lamar Alexander chooses the Internet to announce his presidential candidacy over other mediums.1999: Sirius satellite radio announced..1999: Peer-to-peer file sharing; Napster is launched.

Current Century

- 2001: A feature film is digitally transmitted over a satellite in Europe
- 2003: MySpace, social networking websites launched.
- 2004: The greatest success stories of social networking site in the world, Facebook is launched.
- 2005: YouTube, the video sharing site, is launched.
- 2006: Twitter, micro blogging is now possible.
- 2010: As a part of iPhone 4, face time is launched.

To Think about it

- 1565 –Pencil
- 1770-Eraser
- 1858-Eraser on pencil! 300 years of development

5.1.1 Understanding 2G, 3G, 4G, 4G LTE, and the future 5G

G in these expressions signify Generation. Each succeeding generation is more faster, secure and reliable than its precedent. Most innovative of all these three is the reliability factor. 1G did not use wireless technology until 2G (the second generation) was released later. Major leap in the technology was when the wireless networks became digital from analog. Now the journey has been all uphill from there. 3G stood for faster data transfer speeds a minimum of 200 Kbps, for multi-media usage and remained a long time standard for wireless transmissions.

Till Today we are challenged to get a true 4G connection, which means upwards of a 1Gps (Gigabit per second) transfer rate if you are immobile and at the designated spot.. 4G LTE (Long Term Evolution) comes very close to close to bridging this gap. True 4G on a wide distributed area may not be available until the next generation arrives;5G?

Defining the Standards of the G's

Each of the Generations as defined has standard requirements to meet.

1G – 1G as a term was not widely used until 2G was available. !G was the first generation of cell phone technology. All it was able to do was make simple phone calls.

2G – The second generation of wireless technology. Addition of a few more features to the standard such as simple text messaging, picture messages and MMS.

3G – Third generation set the standards for most of the wireless technology we have come to know and love. Web browsing, email, video downloading, picture sharing, fixed wireless internet, video calls and consequently smartphone technology introduced in the third generation of cell phones. 3G standard requires that the technology should handle around 2Mbps (Megabits per second).

4G – The speed of this technology for wireless communication needs to be in the range 100 Mbps - 1 Gbps per second to be called as 4G. The network resources to support more simultaneous connections on the cell must also be shared as per the 4G standard. After developing fully, 4G could easily surpass the speed of an average wireless broadband Internet connection. Fewer devices could support the full throttle operations of 4G when the technology was first launched. Outside of the transmitted area, 4G phones were working as the 3G phones. When 4G first came on scene, it was simply a little faster than 3G. 4G is not the same as 4G LTE which is more developed to meet the criteria of the standards.

A 4G phone should comply with the standards but finding the network resources to fulfill the true standard was a rarity. We were buying 4G capable devices before the networks were ready and really capable of delivering true 4G to the device. We think that 4G is faster than 3G so we are ready pay the extra price for the extra speed.

4G LTE – Long Term Evolution – When we talk about LTE Advanced, then in fact we will be talking about true 4G wireless technologies because they are the only formats verified by the International Telecommunications Union as True 4G at this point in time.

But forget about that because 5G is coming soon to a phone near you. Then there is XLTE which is a bandwidth charger with a minimum of double the bandwidth of 4G LTE and is available anywhere the AWS spectrum is initiated.

Verizon, T-Mobile and Sprint have all advanced to the LTE technology with each carrier adding their own combination of wireless technologies to enhance the spectrum.

5G – It is rumored that 5G is being tested although the specifications of 5G have not been formally defined. We can expect the new technology to be rolled out around 2020 but in this fast-paced technological world it could be much sooner than that. Seems like a long way away but we are eagerly awaiting 5G at speeds of 1-10Gbps.

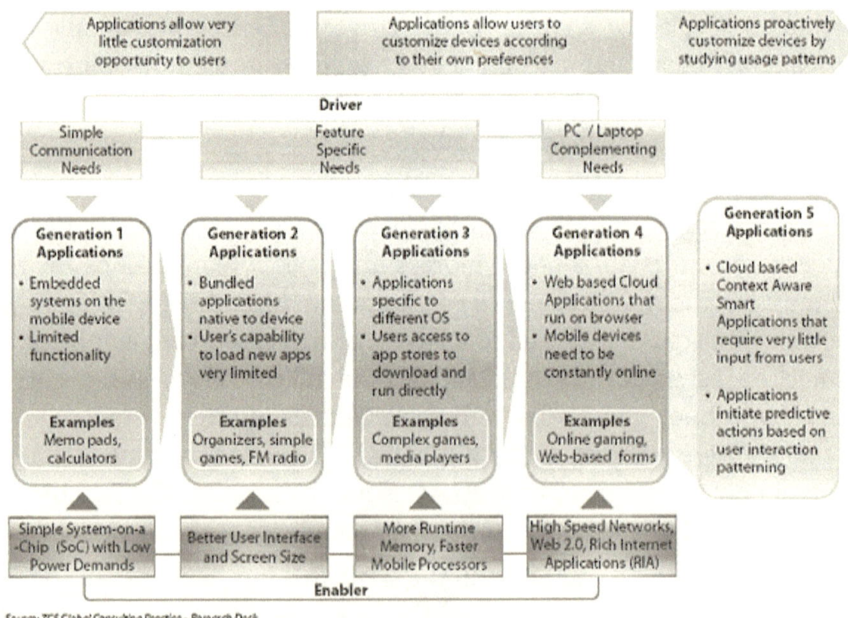

Fig.5.2 Evolution of Mobile Applications.

5.2 Evolution in Transportation Technology

Since the transportation in itself is a wide and varied field for example for civil engineers it would mean Highways, Roads, Ports or Airports for mechanical engineers it might just mean the vehicle of transport Car, Aeroplanes or Hovercraft;similarily other branches may perceive it in terms of carriage of signals only. So with an in ordained task of just accumulating basic knowledge we just delve on very preliminary aspect of transportation and it's evolution over time and to top it up just a brief on the evolution of aviation transport in the US.

The sledge was used to drag heavy loads as it was easy to slide in the snow (7000-4000BC) The next natural stage was the addition of wheels (3000BC)" The wheels on the first wagons were made either from a single piece of wood or from three joined planks; sometimes they turn on the axle, sometimes with it. Speed is not the main characteristic of such a vehicle, as anyone will know who has seen bullock carts on the farm roads of India today"[5]. For greater armor, and far greater speed, two new elements are needed - the horse and a spoke wheel (2000BC). Now with a horse and chariot one could easily move @8 miles p hour as compared to a bullock cart ride@ 2miles per hour. The purpose of the Roman roads (2nd centuryBC-2nd century AD) is the purpose of speed of communicating, so there are post houses lined along the road with fresh horses every 10 miles and lodging for travelers every 25 miles. Because of the straight line roads it results in some very steep hills to cover for haulage purposes these roads are not satisfactory. Person with a horse wagon would prefer an altitude rise less steeper than built by the Roman road engineers. Grand Canal by Chinese so that barges could move from Yangtze to yellow river (3rd century AD-13th century AD).Another technological marvel (to allow passage of vessels between two different levels of canal) is the flash lock and pound lock besides SanMarco lock designed by Great Leonardo da Vinci himself. (10-15th century) Most remarkable is the Canal du Midi, 1681, which joins the Mediterranean to the Atlantic by means of 150 miles of man-made linked waterway linking both Aude and Garonne rivers. At some point this canal descends 206 feet in less than 32 miles; three aqueducts are constructed to carry it over the rivers; a tunnel 180 yards long pierces through a patch of raised ground! The potential of canals is very much self-evident.

Fig.5.3 Canal de mudi.

Britain was the first nation in the 17th century, to construct the first integrated system of waterway transport.

Both eastern and western nations in the late Middle Ages see vast improvements in long-distance travel by sea. China being the leader. While Europeans continued with ocean journeys in long and narrow ships with a single square sail (the long ships of the Vikings as widely known), the Chinese are bent upon improving the design of the junk. A different design helped Chinese general zheng hen conquer Africa. A caravel holding Magellan's crew goes for the first circumnavigation of the globe in 1519-22.

(12th-15th century)

In the middle ages in the 800s, we began paving streets with tar. In the late 13th century we made rockets which could fly. We could cross large oceans in the 15th century by using advanced sailing ships. In the 16th century we started to use horse powered rails made of wood and stone. In 17th century we launched the first oar-propelled submarine. In 1664, we invented the horse-drawn bus. In1672, we could use the first steam-powered car. In the 18th century we used the foot-and-hand-powered carriage. In 1760, we used iron rails In 1769, we trial ran the steam-driven artillery tractor..In 1776, we propelled submarines by the help of screws. In 1783, we launched the first hot air and hydrogen balloons. In 1784, we finally built a steam carriage. In 17 th century the rage for coaches from carriages then to sedan and finally evolved coaches makes interesting reading.

"Etienne makes the journey on their joint behalf and constructs a balloon to be launched at Versailles on September 19, 1783 in the presence of Louis XVI. This time the flying globe or aerostatic sphere (both are contemporary phrases) carries living passengers - a sheep, a cock and a duck. The trio travel more than two miles and land unharmed, except that the cock has been kicked by the sheep."As said in historical records.

Charles acquires 500 lb. of sulphuric acid and 1000 lb. of iron filings to provide enough hydrogen for the balloon. Hydrogen gas is passed for 4 days through lead pipes into a slowly inflating balloon. Cannon is fired to signal the launch At last, on 27th of Aug. In front of an ecstatic crowd on the Champ de Mars the balloon rises rapidly to about 3000 feet.

The hydrogen balloon flies a distance of fifteen miles in about 50 minutes before causing a leak and crashing to the ground near a village. The peasants of the village at the scene of accident are, alarmed by the arrival of this monster from the sky, take to beating the balloon until it seems dead to them. Impressive as these adventures were they used barometers to measure pressure as they were unguided unmanned explorations.

In the 20th century (1900) we built airships. In 1903, we flew motor-driven airplanes and sailed in diesel engine driven canal boats. In 1908, we drove gas engine automobiles. In 1911, we launched diesel engine driven ships. In 1912, we launched liquid-fueled rockets. In 1935, we built DC-3 transport aircrafts. In 1939, we built jet engine-powered aircrafts. In 1942, we launched V2 rockets. In 1947, we had supersonic manned flights. In 1955, we had nuclear-powered submarines. In 1957, we launched a man made satellite into orbit — Sputnik 1, built container ships and flew commercial Boeing 707s. In 1961, we launched the first manned space mission orbiting the Earth. In 1969, we flew Boeing 747 wide body airliners and made the first manned moon landing — Apollo 11.In 1971, we launched the first space station. In 1976, we flew the supersonic concord passenger jet. In 1981, we flew the Space Shuttle. In 1994, the channel tunnel opened.

5.2.1 Early developments in Aviation

- 1903: Historical first flight by Wright Brothers' in North Carolina
- 1916: Air mail service employed by the Army for the first time.

    1920:Earlycommercial flight

    Ford Fokker Trimotor

    Cruising Speed =175 Kph Passengers on board=10-12

    Length of Runway=600m

- 1926 - First Air Commerce Act in the U.S
  - Makes aids to air navigation essential
  - Some authority for traffic rules is in place.
  - It becomes mandatory registration for aircraft providing air services
  - Airmen Certification is made.
- 1927: Services between Miami-La Habana is Pan American World Airways

(Cuba)

1930:

World war II Period

1938:President Roosevelt creates CAA (Civil Aeronautics Authority)

1939-1945: WWII aids aircraft technology boost.

1939-1945: Low cost Airports are created throughout the world to train pilots in scores of numbers

1945-1947: Surplus aircraft Available (specially many DC-3or C-47)

Post War

1945:First control tower equipped with Radar technology.

Douglass DC-6

Cruising Speed =550 kph Passengers on board=45-65

300 built A few are flown even today.

1951:First Commercial Jet (Comet I) launch by the British

  - Length of Airport Runway grows substantially
  - Support equipment is increased in number.

1967-Department of Transportation emerges in the U.S.

  - FAA (Federal Aviation Administration)is formalized.
  - NTSB (National Transportation Safety Board)is created.

Development of Airport in the 60's

(1957-1961)Los Angeles International

(1967)San Francisco International

(1967-1973)New York International

(1970-1990)Development in Twin Engine Aircraft

- Design of large turbofan and super fuel efficient engines
- Airbus and Boeing Introduce long range, reliable twin engine.

Boeing 767-200

Cruising Speed =985 kph Passengers on board=270 Length of Runway=2, 700m Range=7000Km.Payload=200, 000Kg.

Satellite revolution:

1993:GPS launched as better system to ATC

Navigation Satellite trials over Atlantic Ocean

1994: Navigation Satellite trials over Trans Pacific routes

1999:DGPS offers near position approaches on landing

Bombardier RJ 100

Cruising Speed =850 kph Passengers on board=50

Length of Runway=1.400m Range=2000Km.Payload=32, 000Kg.

Development of regional aviation

(1980-1990): Turbojet engines begin successfully replacing old Turboprop technology.

- Turboprop is noisy and risky is the public perception hence this augments the demand for

Turbojet engines.

Very long range Aircraft

Early 2003: Airbus A340-500 and Boeing 777-200ER

Boeing 777-200ER

Cruising Speed =910 kph Passengers on board=300

Length of Runway=3000m Range=13000 Km. (Newyork-Delhi) Payload=32, 000Kg.

Boeing 787 Dream liner

A long Range, mid size, wide body, twin engine jet airliner, most fuel efficient

(20% more efficient than 767),

composite materials in the airframe.

Distinguishing features: electrical flight systems,

Swept wingtips and noise reducing chevrons on its nacelles.

5.2.2 Evolution in Transportation Technology 21st Century

In the 21st century, (2001) we launched the first self-balancing personal transport. In 2004, we operated commercial high-speed Maglev trains and launched the first suborbital space flight — Spaceship One. In 2012, we have now probed and viewed beyond the edge of our solar system with Voyager 1 spacecraft

So, where do the remaining years from this century and the future of transportation now take us to? Back to the moon, to Mars, or to the Jupiter and much beyond? Will we be able probe the greatest depths and heights of the Earth and exceed the greatest speeds and teleport the holographic particle forms? Will there be an sudden end to our inventive and transportation horizons? Or else would l we continue to go where no man or woman has ever dared to go before and beyond?

It is our nature to explore, encompass and conquer the world and the many potential worlds we now appear to know. Curiosity within us explains this unstoppable quest for knowledge and control. Only what limits our creative and new endeavors will be the self-imposed restrain and due to misunderstanding of our laws of the universe. Our compelling desire to unravel the universe leads us onward and on to those new and refined transportation possibilities future holds for us. This is an inspiring reflection upon from where we have come and where we will go to.

So to say these developments were multispatial (air, sea. land) and technologies did not follow or preceded one another but consequential unlike in the technology development studies for evolution of communication.

2000-to Date.

2002: The Segway PT is a two-wheeled, self-balancing, battery-powered electric vehicle invented by Dean Kamen.

2004: A magnetic levitation train, or *maglev* (The Shanghai *Maglev* Train) is a line that operates in Shanghai, China

2010: A personal rapid transit, PODCAR system developed by the British engineering company ULTra Global PRT: ULTra (Urban Light Transit) first public system opened at London's Heathrow Airport in May 2011. It consists of 21 vehicles operating on a 3.9-kilometre (2.4 mi) route connecting Terminal 5 to its business passenger car park, just north of the airport.

5.3 An Introduction to Lear Jets

Fig.5.4 Aircraft, the pioneering Learjet 23

Learjet is instrumental in manufacturing a private, luxury aircraft. First design of Lear was based upon an experimental military aircraft replacing ducted fan turbo shaft engines with turbojet engines This first design was aborted for a Swiss ground-attack fighter aircraft, the FFAP-16 for some reasons. The landing gear and wing with tip fuel tanks of the first Learjet's were little modified by Lear and team from those used by the fighter prototypes.

Come September 19 1968, the company was renamed Lear Jet Industries Inc.

The first Model 25 powered by a Garrett TFE731-2 .This aircraft later on became the successful aircraft Learjet 35.By 1974 the company had outscored every competitor in the field by surpassing the one-million flight hours mark and in 1975 the company produced its 500th jet.In 1976 company increased monthly aircraft production to ten,

The Learjet 28 was made by redesigning Learjet 25with a new type winglet fitted wings. This resulted in an increase in both the performance as the well fuel economy and hence inspired the name "Longhorn";for the short-lived Learjet 28/29 and for few more successful models.

The first prototype for the Model 54/55/56 series was flown in April 1979. In 1983 Model 55 set new records least time to climb records in its category.

The Aerospace Division of Learjet was awarded a contract to produce parts for the Space Shuttle's main engines in 1985,

After purchase by Canadian company Bombardier Aerospace purchased the Learjet Corporation was the marketing the aircraft models as the "Bombardier Learjet Family in1990.In Oct 1990. the Learjet 60 a mid-sized aircraft made its maiden flight, followed by the Learjet 45 in 1995.

Bombardier Learjet launched a brand new aircraft program, the Learjet 85 In 2007; the first FAR Part-25 all-composite business aircraft.

In October 2015 Bombardier announces the cancellation of the Learjet 85 program.

Lear jet 70

Passengers capacity: 7 (standard configuration)

Engine: Honeywell TFE731-40BR

Thrust: 17 kN

Flat rated to: ISA + 23°C, SL

Avionics

- Garmin G5000 with 3 high resolution 14" displays
- Touch screen controllers
- Synthetic Vision System
- Dual Flight Management System
- Graphical flight planning
- Solid state weather radar
- Digital Audio System
- Aircraft position on taxiway
- Data link capabilities

Entertainment

- 7" touch screen display at most of the seats

with Audio/Video control

- Forward 12.1" HD bulkhead monitor
- HD and Ethernet backbone
- Full audio system with hidden trim panel speakers

Range

Maximum range 3, 815 km

(Range with NBAA IFR Reserves, ISA, 4 pax/2 crew. Actual range will be affected

by speed, weather, selected options and other factors.)

Speed   km/h

High-speed cruise 860

Long-range speed 796

Airfield Performance

Takeoff distance (SL, ISA, MTOW) 1, 353 m

Landing distance (SL, ISA, MLW) 811 m

Operating Altitude

Maximum operating altitude 15, 545 m

Initial cruise altitude (MTOW) 13, 716 m

DIMEN SIONS

Exterior

Length 17.1 m

Wingspan 15.5 m

Wing area.7 m2

Height 14 ft 0 in 4.3 m Interior

Cabin length 17 ft 8 in 5.39 m

 (From cockpit divider to most aft cabin

without baggage compartment.)

Cabin width (maximum) 5 ft 1 in 1.56 m

Cabin height 4 ft 11 in 1.50 m

Weights**

Maximum ramp weight 21, 750 lb 9, 866 kg

Maximum takeoff weight 21, 500 lb 9, 752 kg

Maximum landing weight 19, 200 lb 8, 709 kg

Maximum zero fuel weight 16, 000 lb 7, 257 kg

Standard basic operating weight 13, 715 lb 6, 221 kg

Maximum fuel weight 6, 062 lb 2, 750 kg

Maximum payload 2, 285 lb 1, 036 kg

5.4 Five Generations of Jet Fighters

Generational identification of the Jet fighters (and not propeller driven aircraft)is based upon the fact that the terminology of generation shift (change) is held to occur when no further technological innovation could possibly be accommodated into an existing aircraft through mere upgrades and retrospective fitouts.Whenever a major change of jet fighter aircraft is brought about by distinguishing advances in Avionics, Weapon systems and Aircraft design ;we say a generational shift has occurred.

TABLE 5.1 Five Generations of jet Fighters-Distinguishing features

Generation	Avionic system	Weapon system	Range of Attack	Intelligence system	Speed	Jet engine
I E.g., MiG-15, MiG-17, F-86 (1940s-1950s)	Basic avionic system	Cannons, Machine guns, Unguided bombs &Rockets	Visual range only	No Radars or countermeasures for self protection	Subsonic	No afterburner
II E.g., MiG-19, MiG-21, F-104, F-5, Sukhoi-7 (1950s-1960s)	Advanced aerodynamics design	Air- to -Air guided semi active Infrared missiles Radar guided missiles for beyond visual range	Air to air combat within visual range	Radar warning receivers	Supersonic speed in level flights	Advanced engine
III E.g., MiG-25, Mirage-III, F-4Phantom-II (1960s-1970s) Multi-Role fighters	Advanced avionics	Improved Missiles. Semi active guided Missiles like AIM-7, Sparrow. "Look- down, Shoot-down Capability "*.Off-bore targeting*	Aerial engagements now beyond visual range. So you could destroy enemy without visually acquiring them	*Improved Doppler Radar	Maneuverability and Supersonic speeds at different levels	Improvement in engine

(contd.)

IV E.g., MiG-29, Mirage-2000, F-16, Sukhoi-27 (1970s-1990s) Multi-Role capabilities	Head up displays and optimized	Could switch roles between Air- to- Air and Air- to -Ground Combat Mode*	*No distinction between control over air and strike missions.	Fly-by-wire "		
IV and ½ E.g., Dassault Rafale, Gripen, F-16 F, Sukhoi-30MKI (1990s onwards) Improving cost of military economics	More advanced avionics, electronics aerodynamic design	Greater weapons carriage capacity and extended range	Increased use of computer technology to convert it into a network centric battle space control mode for multi role missions.	Stealth fighters "non discoverability on enemy's Radar.Acive electronically screened area (AESA)Radar. Helped fighter plane to act as a limited airborne early waning and control system		Thrust vector controlled engines such as Hornet, Eagle.

(contd.)

V E.g., Dassault Rafale, American F/A-22 Raptor F, Russian Sukhoi-T-50 (Now)	Thrust vectoring and multi sensor integrated avionics	Fighters lethality and Survivability are important consideration for this generation of fighters.	Efficient Software support for e.g., 7 million lines of code on board and 7 million lines	Improved spatial awareness through multi spectral sensors. Pilot can have a 360degree		Supercruise:achieve supersonic speeds without the use of afterburners
V and ½ Multi role with advanced stealth technology			of coding on supporting ground system. Over 100 times more parameters can be controlled as compared to any IVth Genearation. Hence optimum survivability.	view without moving himself or the aircraft.		

(contd.)

Fig.5.5.Dassault Rafale.

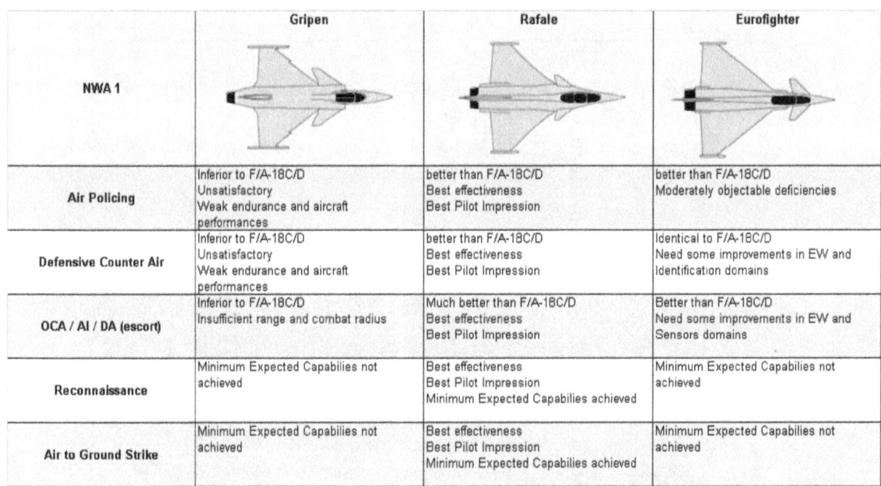

	Gripen	Rafale	Eurofighter
NWA 1			
Air Policing	Inferior to F/A-18C/D Unsatisfactory Weak endurance and aircraft performances	better than F/A-18C/D Best effectiveness Best Pilot Impression	better than F/A-18C/D Moderately objectable deficiencies
Defensive Counter Air	Inferior to F/A-18C/D Unsatisfactory Weak endurance and aircraft performances	better than F/A-18C/D Best effectiveness Best Pilot Impression	Identical to F/A-18C/D Need some improvements in EW and Identification domains
OCA / AI / DA (escort)	Inferior to F/A-18C/D Insufficient range and combat radius	Much better than F/A-18C/D Best effectiveness Best Pilot Impression	Better than F/A-18C/D Need some improvements in EW and Sensors domains
Reconnaissance	Minimum Expected Capabilities not achieved	Best effectiveness Best Pilot Impression Minimum Expected Capabilities achieved	Minimum Expected Capabilities not achieved
Air to Ground Strike	Minimum Expected Capabilities not achieved	Best effectiveness Best Pilot Impression Minimum Expected Capabilities achieved	Minimum Expected Capabilities not achieved

Fig.5.6 Comparison of Modern Fighter planes

Chapter 6

Introduction to any one as a case study:

1. Communication Design
2. Industrial Design
3. IT Integrated Design
4. Textile Design
5. Inter disciplinary Design

6.1 Toyota Case Study

Automotive engineers in U.S were invariably trained to use the problem solving method also called hill-climbing process: each successive solution being a step towards the best possible design at the top of the hill. Since the process moves from one point to another point in the realm of possible designs, we equally refer to it as point-based design. U.S. engineers, authors, and researchers often emphasize "speeding up the iterative loop" by increasing communication feedback among team members through collocation or sometimes computer support or by providing more powerful analytical tools. They emphasize "reducing the number of iterations," by "taking a prototype out of the design cycle," "doing it right the first time," or "freezing specifications early."

Now Imagine what Toyota does, it goes through so many iterations, generating different designs. And Toyota take so long to synthesize the first solution, providing the starting point for modifications unlike others.

Many companies seem to be looking for a design process cookbook, a step-by-step method that, if properly executed, produces a high-quality product quickly and efficiently. But teams seeking to reengineer development processes are often frustrated because rearranging the steps does not offer much improvement.

The principles outlined above are not steps, prescriptions, or recipes. Rather, Toyota chief engineers apply the principles to each design project differently. Design engineers use the principles to develop and evaluate a design process. The key to success is the implementation of ideas as much as the principles themselves. We are still trying to understand all the determining factors, which we hope will lead to guidance on how to implement the system. Future work will address the implementation of these principles in U.S. companies (as well as deepening our understanding of Toyota), but a few issues seem clear.

Basic Principles used by Toyota are given under; but understanding them requires more than we could possibly cover in our syllabus.

Principle 1 — Map the Design Space

Define Feasible Regions

Explore Trade-offs by designing Multiple Alternatives

Principle 2 — Integrate by Intersection

Impose Minimum Constraint

Seek Conceptual Robustness

Principle 3 — Establish Feasibility before Commitment

The Toyota Model

Toyota does not conduct a point-to-point search but follows a version of set-based concurrent engineering

In fact, we increasingly hear of U.S. companies that successfully use set-based approaches. For example, when the aircraft engine division of General Electric wanted to reduce development lead times, it shifted to a set-based strategy in which functional groups together selected feasible solutions, carried them through in parallel until the preferred solutions were verified, and then narrowed the set. The process enabled the development team to avoid much of the rework that might normally occur late in the project and to achieve the aggressive lead-time targets.

1. The Toyota team defines a *set* of solutions at the system level, rather than looking for a single solution.

2. All sets of possible solutions for various subsystems are defined.

3. Possible subsystems are explored in parallel, using design rules, analysis and experiments to characterize a set of possible solutions.

4. Analysis is done to gradually narrow the sets of solutions, converging slowly toward a single solution. In fact, the team uses analysis of

the set of possibilities for subsystems to determine the appropriate specifications to be imposed on those subsystems.

5. Once the single solution for any part of the design is clearly established, it does not change it unless absolutely necessary. In particular, the single solution is *not* changed to gain improvements like the climb the hill process.

"While none of these concepts is new, a set-based philosophy in concurrent engineering extends well beyond the sum of the parts. We argue that the philosophy explains Toyota's unusual practices. By communicating a whole set of possibilities simultaneously and avoiding changes that move outside the set, Toyota can reduce the frequency of communication, eliminate the need for dedication and collocation, and reduce supplier communications for a simpler process structure. The large number of prototypes and alternatives is not a consequence of rapid changes in the design concept, but an effort to explore broad regions of the design space simultaneously. The "ambiguity" in Toyota's specifications to suppliers is actually great precision: Toyota provides only the degree of constraint in which it is confident, avoiding future changes. The philosophy may represent a new model that explains Toyota's success."

Key paradox of the Toyota development process is that, despite its obvious effectiveness, many steps appear extraordinarily inefficient (see Table). There are several aspects of set-based concurrent engineering evident in the approaches of Toyota and its suppliers which can be understood from the table below..

Compared to the competition, Toyota and its suppliers:

- "Explore a larger number of possible concepts in 1/4 or 1/5 scale clay models and expend more resources. Toyota's general manager of styling commented that they "prefer lots of torpedoes to a single sniper bullet."

- Delay fixing body hard points (key dimensions that determine body shape), thereby increasing the design team's uncertainty and decreasing the time for the stamping die designers after the body shape is fully determined. According to Toyota's general manager of body engineering, "The manager's job is to prevent people from making decisions too quickly."

TABLE 6.1 Toyota design process Vs. Other Automobile Manufactures

Area	other Auto majors	Toyota & Suppliers
Number of 1/4 or 1/5 scale Model developed	Japanese OEM 3 to 5	Toyota 5 to 20
Body hard points	U.S.OEM-fixes before full sized clay model avoids changes	Toyota as much as 2cm. design tolerance at first stage of full-sized clay model, fixed at second stage.
Exhaust system specification	Japanese supplier receives hard specs and test data, argues if impossible to meet	Toyota and supplier specifications are approximate targets until second prototype
Number of exhaust system designs prototyped	Japanese Supplier:1	Toyota Supplier:10 to 50
Exhaust systems are relatively simple to design and make and may still be undetermined at the first vehicle prototype stage	Same step by step approach for both Transmission system &Exhaust system	uncertainty is reduced to within 5 percent of the final specification; the subsequent prototype stage reduces the uncertainty to zero just months before production.
Transmissions are very expensive, long lead-time items. For most vehicle programs, the transmission is decided on very early.	Same step by step approach for both Transmission system &Exhaust system	Thus the "transmission gate" is at the vehicle concept stage when uncertainty is reduced to zero years before production. So different approaches to different systems
Cooling fan specification	Japanese OEM provides hard spec at full sized clay model	Toyota supplier:design tolerance-30% at first prototype 5% at second prototype
Number of fans prototyped	Japanese OEM (about supplier):2 to 3	Toyota supplier:average of 4 to 5.30 maximum

- Delay releasing final specifications to suppliers until relatively late in the design process. This conscious decision is based on supplier capability. Toyota sees certain suppliers as less technically capable and provides hard specifications earlier. But Toyota managers report that they provide approximate specifications, or targets, to their more capable suppliers. Approximate targets give the supplier latitude to explore alternatives and are deliberately set to challenge suppliers to excel.

- Develop prototypes of an extraordinary number of different designs for subsystems.

6.2 Design of Mechatronic product with automation software

Modern day products are combinations of various functions say for example the mechanical products are more mechatronical products; combination of mechanical and electronic components. When it comes to an interdisciplinary design project Functions and behavior as a neutral to any branch of a technical stream becomes more and more important. Strong participation of various disciplines has increased the complexity of the integration of mechatronic components during the design Process. Software plays an increasingly important role in the context of modern interdisciplinary product development. When components intensively communicate amongst each other, new functionalities and features are possible. In such scenarios we are speaking of cyber-physical and cybertronic systems. Current research initiatives focus on technological progress with software intensive embedded systems in industrial products. Software allows a number of additional product features at the design stage. This requires an even greater involvement of the software engineering discipline in the design. Therefore it is important to increase the transparency between disciplines.

Now in this case of interdisciplinary product design there is a lack of established, i.e. industrially used methods, processes and IT solutions for such an integrated system/product. So in turn, there is lack of intelligent and networked products and production systems. In the 60s and 70s a development methodology has been proposed in Germany, which is based on functions as a way of describing the product. It was meant for mechanical design and was not based on formal languages at that time.

At the same time many software development methods based on formal languages were developed at an international level (see Fig.6.1).

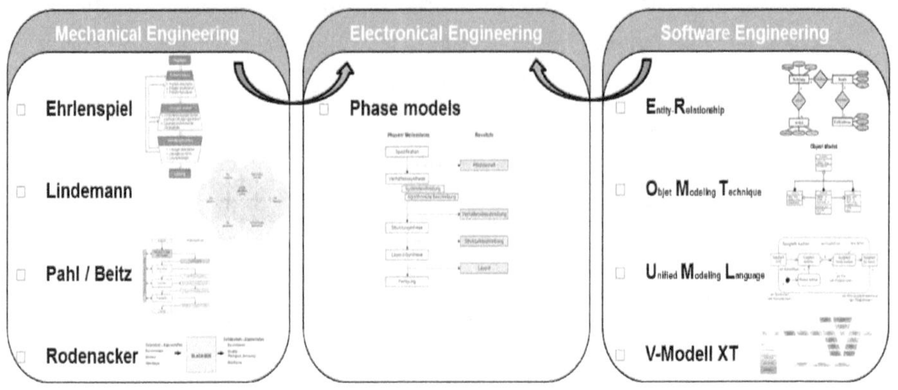

Figure 6.1: Discipline-specific methods and development processes

Since the 50s Systems Engineering (SE) has been defined and used as an interdisciplinary document-driven approach for the development and implementation of complex technical systems in large projects, especially for the U.S. aerospace and military. This approach has been permanently extended from the perspective of software and electronics industry and offers nowadays modeling and simulation support of complex, highly networked systems.

Product Lifecycle Management (PLM) and ERP (Enterprise Resource Planning) concepts that are widely used today does not meet the requirements of the interdisciplinary product and production system development in the early phase of the product design. The data models in use until now are too rigid and the separation in monolithic systems for development (PLM), logistics, personnel and production management (MRP, ERP), customer and supplier integration (CRM, SCM) is obsolete. It no longer facilitates a tracking of complex relationships within the products themselves and their interaction with each other especially in the early phase of the development process.

At the various stages of the Product Development Process, there is already quite a variety of application languages and tools:

- modeling and specification -> SysML (OMG standard), ModellicaML
- modeling and simulation -> Matlab/Simulink, Simscape, Modellica

- discipline-specific modeling and simulation -> (ElectricalCAD) E-CAD, M-CAD, CASE (Computer Aided Software Engineering) and CAE solutions

The interfaces between the three modeling stages listed above are only partially existent and need to be developed and implemented in order to achieve a continuous, consistent design process.

In summary, the current situation can be characterized by a lack of integrative methods, processes and IT solutions – both on the administration side (PLM and ERP) and on the application side – as well as their lack of integration in a consistent modeling approach for requirements (A), functions (F) logical (L) and physical (P) description of interdisciplinary products and production systems. The approach of systems engineering and in particular the model based design could be a future guide to methods, processes and IT solutions for the development of interdisciplinary products and production systems.

6.3 Differences in design of Hardware and Software.

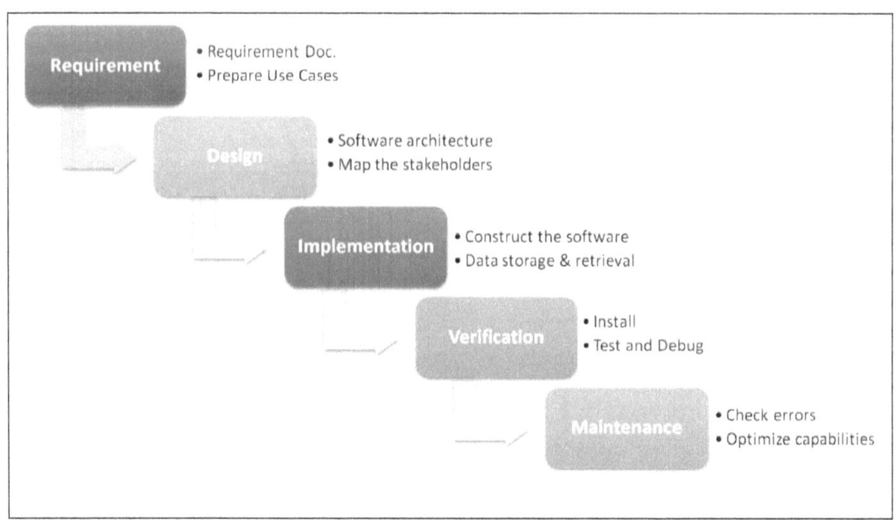

Fig.6.2 A software design process.

TABLE 6.2 Differences between Software and Hardware and so in their design

Hardware	Software
Manufactured	Developed or Engineered
Wear and tear on the usage	No wear and tear
Product life cycle follows a bathtub curve	Software life cycle follows ascending spiked curve
Trend is towards component based construction	Software is custom built too complex to build

Software Engineering – Defined

- (1969) Software engineering is the establishment and use of sound engineering principles in order to obtain economically software that is reliable and works efficiently on real machines

- (IEEE) The application of a systematic, disciplined, quantifiable approach to the development, operation, and maintenance of software; that is, the application of engineering to software"

Software Engineering is a Layered Technology

Process, Methods, and Tools

- Process
  - Provides the glue that holds the layers together; enables rational and timely development; provides a framework for effective delivery of technology; forms the basis for management; provides the context for technical methods, work products, milestones, quality measures, and change management

- Methods
  - Provides the technical "how to" for building software; rely on a set of basic principles; encompass a broad array of tasks; include modeling activities

- Tools

Provide automated or semi-automated support for the process and methods (i.e., CASE tools) What is a Software Process?

- "(SEI) A set of activities, methods, practices, and transformations that people use to develop and maintain software and the associated products (e.g., project plans, design documents, code, test cases, and user manuals)

- As an organization matures, the software process becomes better defined and more consistently implemented throughout the organization

- Software process maturity is the extent to which a specific process is explicitly defined, managed, measured, controlled, and effective"

6.4 Software designing Models- a study

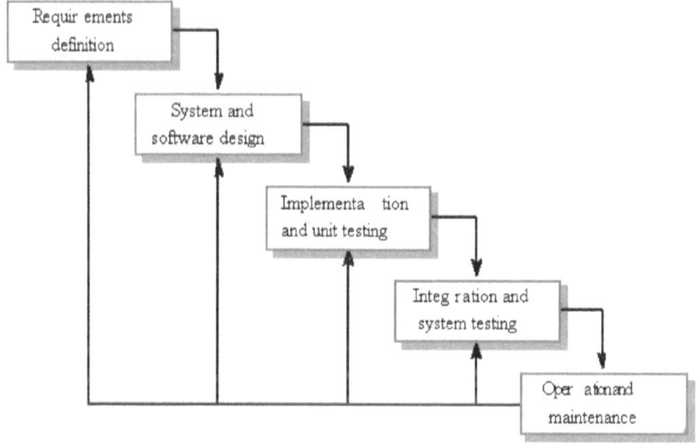

Fig.6.3 SDLC Waterfall Model

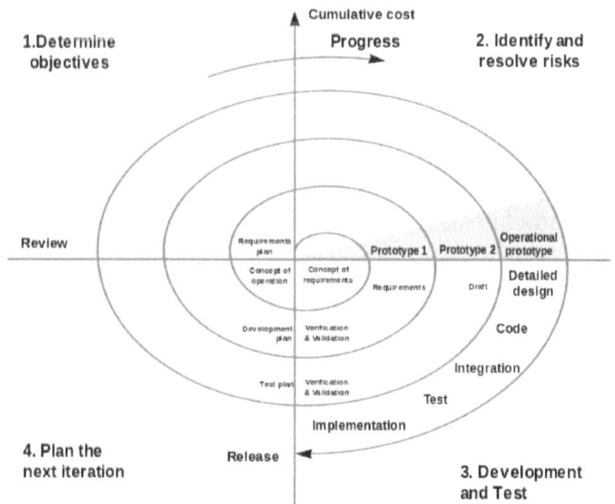

Fig.6.4 Spiral Model for Software Designing

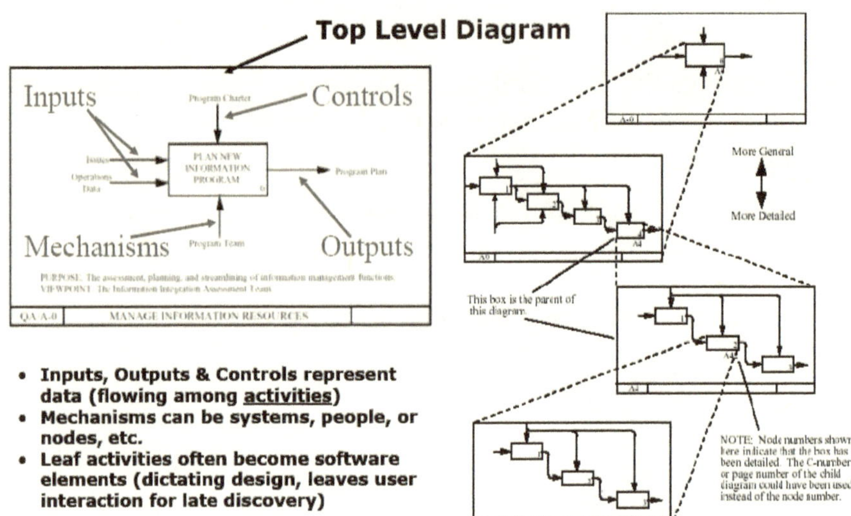

- **Inputs, Outputs & Controls represent data (flowing among _activities_)**
- **Mechanisms can be systems, people, or nodes, etc.**
- **Leaf activities often become software elements (dictating design, leaves user interaction for late discovery)**

Fig.6.5 A software develpoment perspective(view)

6.5 DIY Design it yourself Project

- Students are being asked to design a simple device to crush aluminium cam. Four solutions. six criteria and assigned weights are:
- safety: 30% (30 points)
- durability strength: 10% (10 points)
- facts if used: 20% (20 points)
- use of standard parts: 10% (10 points)
- portability: 20% (20 points)
- cor: 10% (10 point)
- FOUR ALTERNATIVE SOLUTION
- a spring loaded crusher
- a foot operated device
- a gravity powered dead weight crusher
- an arm powered lever arm crusher

6.6 Systems approach as analysis-synthesis for product design.

For the sake of familiarity let us consider an engineered system, an automobile. When we drive it, it works as a single product. A trip to the garage

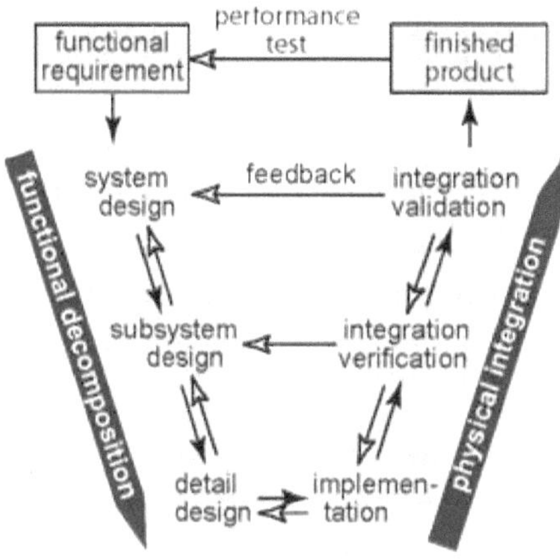

Fig.6.6 Vee Model for product designing.

will convince us that it is a system composed of many interrelated subsystems: transmission, ignition, steering, power train,, braking, lubricating, suspension, and more. Each of these subsystems is again made up of many parts, for instance the clutch, lever, and gear box for transmission. The gear box an assembly in itself consists of many components, and so on unto the nuts and bolts.

Engineers design cars. The systems approach used by them is illustrated in many textbooks by the *Vee* model (for example); originally due to Forsberg and Mooz. The downward stroke of the *Vee* represents functional analysis, the upward stroke physical synthesis as shown in Fig.6.6

Consider this: An Airbus A380 has an approximate 4 million parts, with 2.5 million part numbers produced by 1, 500 companies from 30 countries around the world, including 800 companies from the United States. Now think of system engineering design or think of 4000 engineers designing at the concurrently for an Airbus!

Engineers do not design a car by starting out with a bunch of nuts and bolts. They think of car as a whole. Obviously, the car does not yet exist. Thus they start with a conception of the intended car, more specifically, a set of *functional requirements* for it: what it is supposed to do, what performances are expected of it. For example how many passengers it will carry? what range of speed will it move?etc., To identify satisfactory functional requirements is usually an important and difficult task, add to it the complexity of the system. We will study it later in the context of systems engineering.

They proceed to functional analysis after a satisfactory conception of the intended car. Functional decomposition of the conceived car is done into functional subsystems with proper interfaces, e.g., a subsystem for driving, a subsystem *for* power and a subsystem *for* transmission, and how the three functional subsystems are to work together. Further subsystem is analyzed into its interrelated components, until they get manageable parts that can be specified to the last details for eg., a gear, a lever etc., Now they are at the bottom of the *Vee*. Turning the corner of the *Vee*, the thousands of parts are manufactured to specifications. They are then tested, brought together, and assembled into larger and larger subsystems, finally into a car ready for test drive. We reach back at the top of the *Vee* in the process but this time we have a concrete system – a real car – instead of a mere conception of it.

Functional decomposition, detailed design, and physical assembly helps engineers know and specify the whole, its parts, and their interrelations clearly and at every compositional scale. Subsystems at intermediate levels are important for managing complex systems as they enable engineers to introduce complex details one step at a time.

In short, the *systems approach* integrates analysis and synthesis. It is most effective in treating complex phenomena, for it:

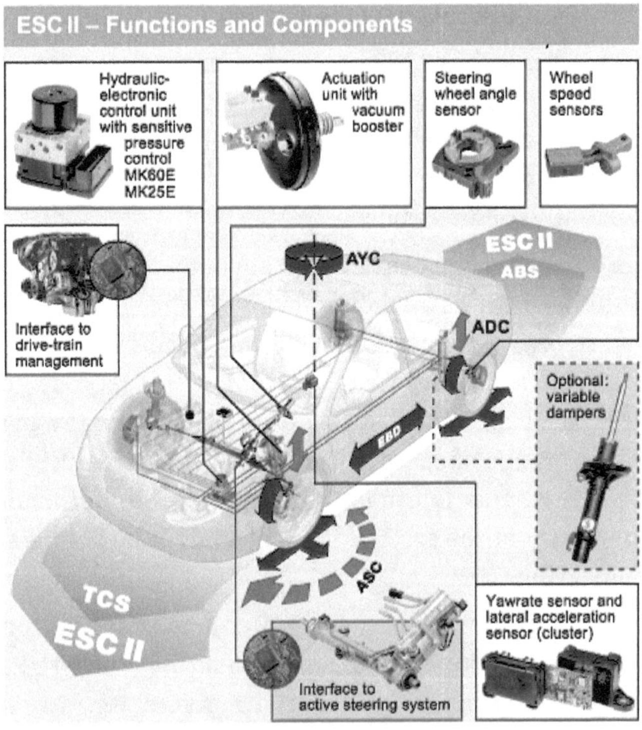

Fig. 6.7A systems and functions approach for a car.

Includes both the holistic and modular views

- handles the arching features of the whole system
- analyzes it into parts with proper interfaces
- synthesizes knowledge about the parts to understand the whole

- understands the system and its details at many levels

  - decomposes a system into subsystems, and so on, to the last details
  - view at different levels so the complexity is reduced.
  - Hides abstracted information to focus on a task:
  - Simplification of the system by treating its parts as black boxes except their interface with other parts
  - Information hiding is not discarding it; a black box can be opened at will
  - makes a complex system more tractable
  - a part can be studied or designed with minimal interference from other parts
  - controls damage and improves safety
  - confines most effects of a defect within a subsystem, preventing a system-wide collapse

The systems approach was first applied to natural science before engineering used systems approach in which one analyzes the details of parts in order to know the whole. For example systems biology, which aspires to study organisms as a whole. The idea appeared in the 1960s, but lied dormant because the properties of the biological constituents remained in the dark. In the meantime biologists analyzed organisms into organs and tissues and cells and molecules. Recently molecular biologists have deciphered the genome of many organisms, including humans. They have now reached at the bottom of the Vee and hence gained tremendous amount of knowledge. Yet they found it falls far short for understanding organisms.

Following analysis they turn to synthesis and investigate how genes function in cells and organisms as a whole. As they do it, systems biology comes alive. Several universities around the globe, including Harvard, are establishing departments of systems biology. Four decades of analyzing organic constituents turn systems biology from philosophy (complex) into science (easy to understand).

6.7 Design of a Special Pump for High Pressure High Temperature Fluids

need- rationalise manufacturing range to reduce cost

- reliability of pump convenient to varied and changing needs of their customers

Clients/customer wanted- reliable

- Convenient

- robust

- Standardised range

(High level and general objectives which were to be investigated further- cracking and leakages due to stresses caused by thermal expansions of pipes

...This is the main problem to which reliability, robustness were aimed

Investigating convenient objective- two sub objectives inlet and outlet ports of pump should always be in line to avoid thermal expansion, such a system with small base size and modular dimensioning of alternating component would facilitate installation and replacement of the pump

Reliable	Convenient
Robust	Easy to install & replace
Resistant to external mechanical stresses	Occupy minimum space
Unaffected by thermal expansion of pipes	Standardize range

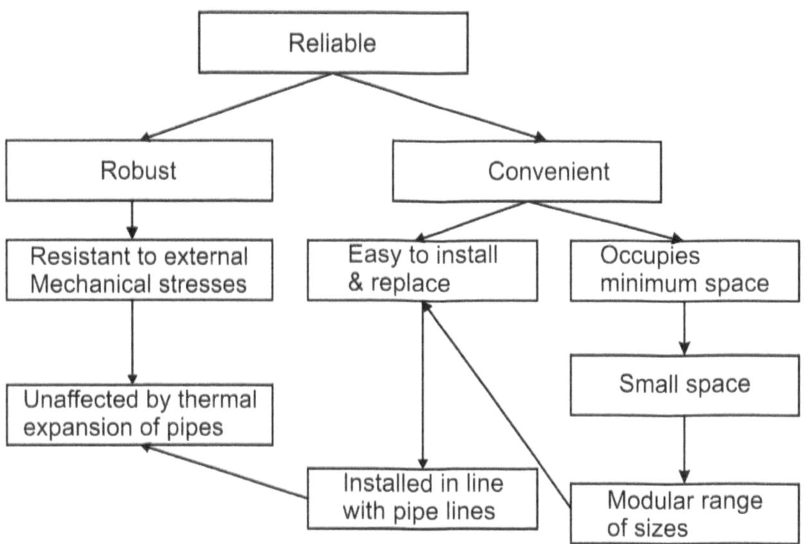

Fig.6.7 Objective tree for a special pump.

6.8 Textile Case

Velcro, the "zipper less zipper," is very common and exists on a variety of products from children's shoes to laptop bags to blood pressure gauges to airplane flotation devices. Velcro is synonymous with the hook-and-loop style of binding, it's actually the name of the company that invented the textile technology and now has many imitators using their patent.

Inspired by nature

Swiss engineer and amateur mountaineer George de Mestral went hiking in the woods with his dog on coming back to his home, he explored the burrs (small piece of material remaining attached to clothes) that clung to his clothes.He was acquiring nature's lessons to the mankind (valid in various other cases)and pondered if such an idea could be useful for commercial application. He studied a burr under a microscope only to discover that they were covered in tiny hooks, which allowed them to grab onto clothes and fur that brushed in passing. Believe this after more than eight years of research work, he created what is known today as Velcro, a combination of the words "velvet" and "crochet." Just made up of two strips of fabric, one covered in thousands of tiny hooks and the other with thousands of tiny loops, the materials gripped together firmly while still allowing easy release see in fig..

Fig.6.9 Velcro uses tiny hooks and loops to bind the two strips. Velcro is a combination of the words 'velvet' and 'crochet.'

As common to several cases de Mestral's invention became the source of much ridicule early in its inception, his perseverance allowed him to perfect the hook-and-loop technology for commercial use. He patented Velcro in 1955, he helped give his company a competitive edge over other would-be imitators, as evidenced by his company reaching the point selling more than 60 million yards of Velcro per year during his lifetime.

Velcro's popularity

Many sources attribute the creation of Velcro to NASA, though this claim is false. Though NASA cannot claim for the material's invention, the space agency's use of the product did lead to Velcro's popularity in all walks of life. Apollo astronauts used Velcro to secure all manner of devices in space for easy retrieval and being lightweight in 1960's.

In spite of the increased awareness created through NASA's use of Velcro, the material could be made only in few colors and often looked misfit on the dress. Starting in 1968 and on into the 1980s due to the simple lack of aesthetic appeal, Velcro was used only with athletic equipment. shoe companies like Puma, Adidas and Reebok integrated Velcro straps onto children's shoes.

At this time, the patent on the hook-and-loop technology had expired and many imitators began to crop up throughout the world. Many of them were low-quality and cheap versions, which forced Velcro to begin a lifelong battle of maintaining the integrity of its product's name to prevent it from becoming a mere generic term, much like aspirin, which was originally a brand name.

Velcro gained popularity in many new styles of use when, a 1984 interview between David Letterman and Velcro's USA director of industrial sales ended in Letterman jumping off a trampoline onto a wall while in a Velcro suit. This prompted many companies to find innovative and versatile methods of using Velcro, from attaching electronic devices to car seats to toys with Velcro materials for catching balls, medical devices and the list goes on and on.

Noise it makes while unbinding and dust removal off Velcro is still a Problem!

Velcro made significant headway in the industry in 2004, by gaining the U.S. Army as a client. The hook-and-loop material became used on the Army Combat Uniform, a lighter version of their original battle attire. However, soldiers disliked the material and caused much uproar through complaints over the noise it created and how the fabrics collected dust. After an internal inspection of these claims, the Army moved away from the use of Velcro to instead rely on buttons.

Sample Questions and Answers

Multiple Choice

1) Which activity would an engineer do, but not a scientist?

    a. record measurements

    b. make observations

    c. draw conclusions

    d. build prototypes*

2) An engineer notices that a bicycle helmet has a flaw in its design. The chin strap separates from the helmet easily, which may cause injuries. What step should the engineer take next to improve the design of the helmet?

    a. draw block diagrams for several new helmets

    b. identify design constraints for bicycle helmets*

    c. build models of several new helmets

    d. gather information about bicycle helmets

3) Which statement describes one way that all types of engineers are similar?

    a. All engineers need to take math classes.*

    b. All engineers make the same salary.

    c. All engineers take the same classes.

    d. All engineers design new products.

4) Why can it take engineers many years to develop a product to be sold in stores?

    a. Engineers test many designs for flaws.*

    b. Engineers wait for scientists to make new discoveries.

    c. Engineers work alone to develop products.

    d. Engineers take their time to apply the scientific method.

5) Which statement about block diagrams is NOT correct?

    a. Block diagrams allow engineers to see if the product actually works.*

    b. Block diagrams can be used as instructions for building prototypes.

    c. Block diagrams illustrate how the parts of a product fit together.

    d. Block diagrams may be developed by teams of engineers.

Short Answer

6) Draw a block diagram of a desktop computer. Be sure to include and label the following parts:

    a. Monitor (screen)

    b. CPU (central processing unit)

    c. Keyboard

    d. Mouse

    e. Speakers

7) Complete the statement "I want to become an engineer to design…" in as much detail as you can.

8) Choose an engineering company that you would someday consider working for.

From the information found on their Web site, write a one-page paper on the engineering problems that they are solving and how they are addressing their customer's needs.

9) Find three textbooks that introduce the engineering design process. Copy the steps in the process from each textbook. Compare with the six steps in our chapter.

10) Interview an engineer working in your chosen field of study, describe in a one page paper what steps of the design process he/she is engaged with in their job.

11) Select a specific discipline of engineering and list at least 20 different companies And/or government agencies that utilize engineers from this field.

12) Choose a product that was most likely designed by an engineer in your chosen Field of study. Identify what problem this product solved, what

constraints were applied to its design, and what criteria were most likely used to evaluate this Design.

13) Choose one of the products from the list below and note key features and functions for the product as produced today. Then, go back one generation (18–25 years) to family, relatives, or friends and ask them to describe the key features and functions of the same product as produced at that time. Note changes and improvements and prepare a brief report.

   (a) Toaster

   (b) Electric coffeemaker

   (c) Color television

   (d) Landline telephone

   (e) Cookware (pots, pans, skillets)

   (f) Vacuum cleaner

   (g) Microwave oven

14) Choose a device or product that you believe can be improved upon. Answer the following questions: 1) what do you already know about this device/product;

2) What do you think you know about this device/product; and (3) what do you need to know about this device/product? Based on your responses to these questions conduct research to confirm or reveal what you know and don't know. Write a short report to summarize your findings. Use proper citation methods to list your research sources.

15) Choose one of the following topics (or one suggested by your instructor) and write a paper that discusses technological changes that have occurred in this area in the past 15 years. Include commentary on the social and environmental impact of the changes and on new problems that may have arisen because of the changes.

   (a) Passenger automobiles

   (b) Electric power-generating plants

   (c) Computer graphics

   (d) Heart surgery

   (e) Heating systems (furnaces)

(f) Microprocessors

(g) Water treatment

(h) Road paving (both concrete and asphalt)

(i) Composite materials

(j) Robotics

(k) Air-conditioning

16) The following list of potential design projects can be addressed by following the six-step design process discussed in the chapter. A team approach to a proposed solution, with three or four members on each team, is recommended. Develop a report and oral presentation as directed by your instructor.

- A device to prevent the theft of helmets left on motorcycles
- An improved rack for carrying packages or books on a motorcycle or bicycle
- A child's seat for a motorcycle or bicycle
- A storage system for a cell phone in a car (including charger)
- An improved wall outlet
- A beverage holder for a card table
- A better rural mailbox
- An improved automobile traffic pattern on campus
- Improved bicycle brakes
- A campus transit system
- Improved pedestrian crossings at busy intersections
- Improved parking facilities in and around campus
- A device to attach to a paint can for pouring
- An improved soap dispenser
- A shoestring fastener to replace the knot
- A better jar opener
- A system or device to improve efficiency of limited closet space
- A shoe transporter and store
- A pen and pencil holder for college students

- A rack for mounting electric fans in dormitory windows
- A device to pit fruit without damage
- An automatic device for selectively admitting and releasing pets through an Auxiliary door
- Ramps or other facilities for handicapped students

17) What is the smallest functional designed object you have heard of? The largest?

18) In a bicycle (or thermostat or other product of your choice), give an example of a standard part, a special purpose part, a special purpose subassembly, a standard module, and a component.?

19) Have you ever used the basic guided iteration process to solve a design problem? If so, describe your experience, noting especially how you implemented each of the steps

20) Have you ever used the basic guided iteration process to solve a problem other than a design problem? If so, describe your experience, noting especially how you implemented each of the steps.

21) Go to your library and find at least one book on design for manufacturing and read its introductory chapter and the preface. Report back to the class on how it is the same and how it is different from this chapter.

22) Identify the stages that constitute "product development" as defined in this chapter.

23) What are the disadvantages to the sequential mode of operation depicted in design steps?

Questions (RGPV)

1. a) Describe four challenges of Design? What is the importance of Engineering Design process?

Sample Answer.

The main challenge faced in designing is an attempt to analyze the problem beforehand and in an abstract isolation from solution concepts; although we can make some progress from problem to subproblem and from sub solutions to solution finally; the modern system requirements requires a a symmetrical and commutative relationship between the problem and the solution, and between sub-problems and sub-solutions. Co-evolution of problem and solution is essentially captured nature of the design process, in which the understanding of the problem and of the solution develop together (for example look at the Vee Model P.). Consistent transfer of the designer's attention backwards and forwards between the problem space (left-hand side of the model) and the solution space (right-hand side of the model) is rewarding. This kind of model also ensures that there is an expected pattern of progression in the design process, from the given problem to a proposed solution. Therefore there is a certain be a general clockwise direction of movement in the model, from top left around to top right, but with substantial amounts of iteration; swinging to-and-fro between problem and solution, sub-problems and sub-solutions.

This process of swinging back and forth and accumulated iterations gives rise to what is now well known also as the "Paradox of Design."

Second challenge is The Paradox of Technology.

Technology us we understand offers the potential to make life easier and more enjoyable; each new technology providing increased benefits to the humankind. With added complexities our difficulty and frustration also increase with technology. This is the "The Paradox of Technology" Technological advances increases the problem of design. Let us cite the example of a wristwatch to prove our point. Few decades ago, watches were very simple. All you were required to do was set up the time and wind the watch for running continuously. The standard control used was only the stem: a knob at the side of the watch. Turning the knob would wind the spring that provided power to the watch to run 24 hours. Pulling out the knob and turning it rotated the hands. The operations were so easy to learn and easy to operate. There was a perceptible relationship between the turning of the knob and the resultant turning of the wrist watch hands. The design even cared for human error. In its normal position, turning the stem wound the mainspring of the clock. The stem had to be pulled before it would engage the gears for setting the time. Accidental turns of the stem did no harm. A fine example of customer centric design.

Watches in olden times were expensive like jewellery manufactured manually. They were sold in jewellery stores. With the introduction of digital technology after some time, the cost of watches decreased rapidly, while their accuracy and reliability increased. Watches became tools, available in a wide variety of styles and shapes and with an ever-increasing number of functions. Watches were sold everywhere, from local shops to sporting goods stores to electronic stores. Moreover, accurate clocks were incorporated in many appliances, from phones to musical keyboards: many people no longer felt the need to wear a watch. Watches became inexpensive enough that the average person could own multiple watches. They became fashion accessories, where one changed the watch with each change in activity and each change of clothes.

In the modern digital watch, instead of winding the spring, we change the battery, or in the case of a solar-powered watch, ensure that it gets its weekly dose of light. The technology has allowed more functions: the watch can give the day of the week, the month, and the year; it can act as a stopwatch (which itself has several functions), a countdown timer, and an alarm clock (or two); it has the ability to show the time for different time zones; it can act as a counter and even as a calculator. Watch has now several functions.. It even has a radio receiver to allow it to set its time with official time stations around the world. Even so, it is far less complex than many that are available. Some watches have built-in compasses and barometers, accelerometers, and temperature gauges. Some have GPS and Internet receivers so they can display the weather and news, e-mail messages, and the latest from social networks. Some have built-in cameras. Some work with buttons, knobs, motion, or speech. Some detect gestures. The watch is no longer just an instrument for telling time: it has become a platform for enhancing multiple activities and lifestyles.

The added functions cause problems: How can all these functions fit into a small, wearable size? There are no easy answers. Many people have solved the problem by not using a watch. They use their phone instead. A cell phone performs all the functions much better than the tiny watch, while also displaying the time.

Third challenge is Paradox of desire and requirements.

"Now imagine a future where instead of the phone replacing the watch, the two will merge, perhaps worn on the wrist, perhaps on the head like glasses, complete with display screen. The phone, watch, and components of a computer will all form one unit. We will have flexible displays that show only a tiny amount of information in their normal state, but that can unroll to considerable size. Projectors will be so small and light that they can be built into watches or phones (or perhaps rings and other jewellery), projecting their images onto any convenient surface. Or perhaps our devices won't have displays, but will quietly whisper the results into our ears, or simply use whatever display happens to be available: the display in the seatback of cars or airplanes, hotel room televisions, whatever is nearby. The devices will be able to do many useful things, but I fear they will also frustrate: so many things to control, so little space for controls or signifiers. The obvious solution is to use exotic gestures or spoken commands, but how will we learn, and then remember, them? As I discuss later, the best solution is for there to be agreed upon standards, so we need learn the controls only once. But as I also discuss, agreeing upon these is a complex process, with many competing forces hindering rapid resolution. We will see."

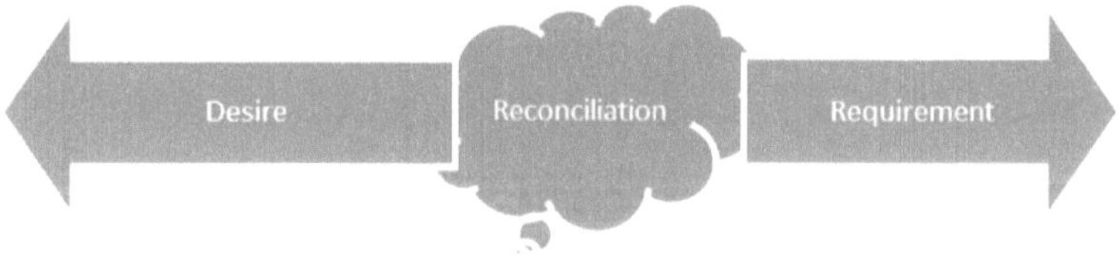

It is design omnipotent task of reconciling between the desires and requirements as in fig. above ;also within his limits, resources and as per schedule.

Last but not the least challenge is CostVs.ScheduleVs.quality performance requirements.

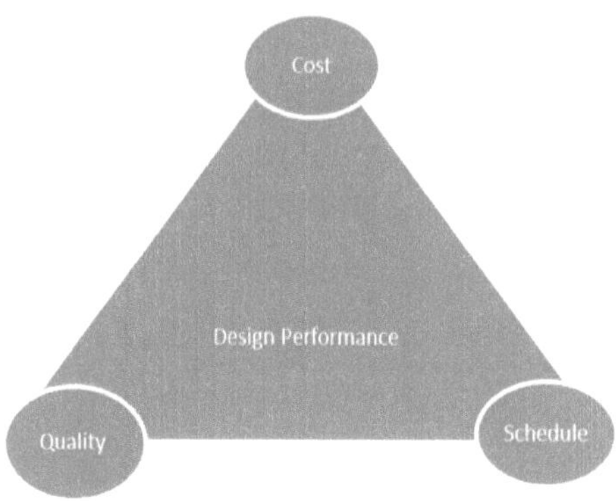

Fig. Cost Vs. Schedule Vs. quality performance

Design performance triangle and at the ends of the triangle are cost, schedule and quality all important parameters of design performance.All the corners of the triangle are desirable but it is possible to have any one only at the cost of the other for example you may increase the quality of a product but only at an increased cost and vice versa.

b) Comment on the importance of Animation and film Design in our daily life as an individual and as a citizen of India.

Sample Answer: Broadcasting and display of moving visual media

Video

- resolution of display
- how are they refreshed
- 3D rido systems

Variety of medium – radio broadcast, tapes, DVDs, Computer files etc.

(CRT)

VTR- Video tape recorder ($50, 000 in 1956 to $300 per hour reel 1971 sony VCR)

Digital technology replaced Analog techbology

DVD in1997

Blue ray disc in 2006

Computer technology allowed personal computers to capture, store, edit and transmit digital video, reducing cost of video production allowing program makers and broadcasters to move to tapeless production.

Digital broadcasting, Digital TV

As of 2015, increasing use of high resolution video cameras with improved dynamic range and colour gamut's and high dynamic range in intermediate data formats with improved colour depth modern digital video technology is converging with digital film technology

Frame rate- PAL, SECAM -25 frames/s NTSC- 29.97/s

Interlaced vs progressive – PAL576 (total no. Of horiz/60

PAC 516/50

516- total no. of horizontal scan lines

50- 50 fields (half frames)/sec)I-Interlaced

In progressive scan system, each refresh period updates all scan lines in each frame in sequence.

Aspect Ratio:

Traditional television screen is 4:3 or about 1.33:1

HDTV 16:9 or about 1.78:1

Full 35mm film with soundtrack (Academy ratio) is 1.375:1

View video of mobile: vertical video

Colour space & bits per pixel

Amount of data is reduced by chroma subsampling; it reduces the no. of distinct points at which colour changes

Video quality

Video quality can be measured with formal metrics like PSNR (peak signal to noise ratio) or subjective video quality using expert observation

Video compression method

- Temporal redundancy

- Spatial redundancy

MPEG -2 (DVD, Blue ray, satellite TV) and MPEG-4- mobile phones, AVCHD (Advanced video coding High Definition), Internet

Stereoscopic

1) Two channels

- Right channel right eye

- Left channel left change

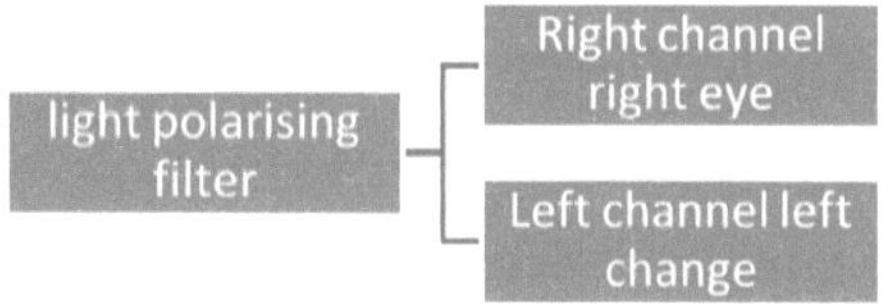

light polarising filter is eye glasses with matching polarization filters.

2) One channel with two overlaid colour loaded layers.

3) one channel with alternating left & right frames for corresponding eye, using LCD shutter glasses that read frame sign from VGA display data to alternatively block image to each eye. so ... eye see the appropriate form e.g. virtual reality

Video formats (refresh rate, display resolution, colour space)

Analogue video- Svideo, VGA, component video (luminance, brightness, chrominance)

Digital Video - DVL, HDMI (High Definition multimedia Interface)

Transportation – analog or digital

- Coaxial cables, dual coaxial cable system

Using progressive scan (SDI serial Digital Interface)

Video display

- CRT
- LED
- Plasma display panel (PDP)
- LCD
- OLED
- Laser TV (forthcoming)
- Surface conduct electron emitter display (SED) (experimental)
- carbon nanotubes (experiment)
- Quantum dot display (experiment)
- Digital Micro shutter Display (DMD) (LED, LCD life 25,000-40,000 hrs)

In 2007 OLED were created to sustain 400 cd/m3 of …. For over 198,000 hrs for green OLEDs 62,000 hrs blue OLED's

Emissive electroluminescent layer (organic compounds) emits light in response to an electric current-AMOLED

Above is of two types-

1. passive matrixworks without backlight (PMOLED)

2. AMOLED

(After 1000 hrs blue luminance degraded by 12% reduced by 7% 14,000 hrs-five years/8 hrs/day)

A multiplexed display

- Fever writes (often fewer wires)
- Simpler driving electronics can be used
- Reduced cost
- Reduced power consumption

The French may have pioneered the theory of moving image, the British might be forerunners, and India would master the art one day and soon to develop largest and biggest movie industry in the world

Silent era (up to 1931)

ArdechirIrani – Alamara

In 1990's

- Advanced special effects
- Dolby digital sound
- Changes in choreography
- Screenplay
- And off course International appeal

Mahabharata, Ramayana, Indian Epic

Multilingual

Digital and celluloid and so on..here the student is expected to pick up from the above or look for and answer to question 1) Comment on the importance of Animation and film Design in our daily life as an individual and as a citizen of India. (RGPV 2015)

The Importance of Animation and Film Design to us (we are a stakeholder here)

- As a engineering student the technology and design space the movies and animation provide.
- Has an 'edutainment' (education plus entertainment) value for us.
- A very popular medium of social education and good living.
- This industry gives employment to millions of people in India.
- An act of cultural expression and national values.
- Lot many interdisciplinary technology growth for like biosciences, manufacturing and electronics of AMOLED.
- Gaming platforms and game design ;use of virtual reality is another extreme technology.
- YouTube, DTH, Video on demand, Video conferencing etc.,form another spectrum of the technology.

The list is endless to be covered in brief especially if you are an avid movie buff!

2. a) What is a problem- solving methodology? Describe its various steps.

   b) What is design paradox? Discuss the various considerations of a good design.

3. a) Discuss about various social issues involved in any design process in Indian context.

   b) Describe total life cycle of a product. How products can achieve its performance requirements?

4. a) What is the smallest functional designed object you have heard of? The largest?

Sample Answer: The powers of ten are commonly used in physics and information technology. They are practical shorthand for very large or very small numbers.

Power of ten	Number	Symbol
10^{-12}	0.000000000001	p (pico)
10^{-9}	0.000000001	n (nano)
10^{-6}	0.000001	m (micro)
10^{-3}	0.001	m (milli)
10^{-2}	0.01	
10^{-1}	0.1	
10_0	1	
10_1	10	
10_2	100	
10_3	1000	k (kilo)
10_6	1 000 000	M (mega)
10_9	1 000 000 000	G (giga)
10^{12}	1 000 000 000 000	T (tera)
10^{15}	1 000 000 000 000 000	P (peta)

We compare the big and the small on a scale we understand. There may be several example from pin to aero planes from inkjet nozzles to space shuttle system from cycle to valodromes from cars to 2.1million ton ships from paper holder to goods carriage trains from windows to golf courses from home network to worldwide web....the choice is yours.But for a certain reason of understanding the scale of complexity involved, the abstraction required,the innovation needed, the inspiration to be drawn from science and nature. We suggest this case.

Name	Building Nanoscale and Micro scale 3-D structures	Large Hadron Collider(LHC)
Purpose of design	Self Assembling Micro and Nanoscale 3-D structures because at this size it cannot be done manually.	Disassembling Atoms to quarks and protons
Size	Nanometer Scale Atom: 10-9 m	Kilometer Scale(27 km in circumference) 103 m
Principle of Design	Compacting shapes so at this scale they would automatically create 3-D structures	Mass-Energy Transfer
Energy Used	Minimal	Design energy of 7.2 x1012 Volts(7.2TeV)per proton beam
Constraints	Our work is inspired by and should mimic replication of viruses	squeeze energy into a space about a million million times smaller than a mosquito
Use to society	"there is a need in medicine to create smart particles that can target specific tumors, specific disease, without delivering drugs to the rest of the body, which limits side effects. "plus wankel rotary engines and many more to be known..	Understanding the beginning of the Universe, discovering Higgs Boson and so on and on..

4 b) In a bicycle. give an example of a standard part a special purpose part a special purpose subassembly, a standard module and a component

5. a) What is the initial process of the design? Describe its activities. Suggest the techniques or the mechanisms which comes in your mind those can bring the improvement in the performance of a Bullock Cart used In our villages.

Sample Answer: Initial Process of Design Include the stages before the designing process begin

Research steps (1-6) (As from our book P.)

10) Identify Needs

  – What's the problem?

Underlying problem the customer wants to solve – for example people didn't use toothbrush for brushing their teeth until 1938 they used other arrangements for cleaning teeth like *neem* and *jamun* sticks were chewed and used as brush ultimately led to toothbrushes and using several tons of nylon for that purpose.

11) Information phase

  – What exists?

12) Stakeholder phase

  – what's wanted and who wants it?

Who are the stakeholders for this books you, your parents because they purchase, publishers for their interest in publishing. Authors for their penchant to write, Teachers using this books for reference, educational quality if this turns out to be good and useful book and so on a very broad list has to be thought out..

13) Planning/operational Research

  – What's realistic? What limit us?

14) Hazard Analysis

  – What's safe (what can go wrong?)

15) Specifications

  – What's required?

NEED (improve bullock cart)

Durable (more than 25 years)

Fossil fuel to be replaced with animal energy

For poor road connectivity and time is limiting factor

Increase agricultural productivity with minimised loss

Environment friendly, pollution free

Fabricated bullock cart would generate rural employment

Trees and bamboos would be replaced with steel tubes

ADVANTAGE

Bullock cart are made of steel, mostly welded, so rigid

Design and configuration are simple, so it is manufacture by rural fabricator easily

Very less weight/capacity ratio about 280 kg for 2 ton capacity, 210kg for 0.5 ton

Centre of gravity is low just above the top of mid axle, to have better balance square or rectangular tube section have been used to provide better load bearing properties to the carts

The cart can be modified to suit any type of use

Mechanism of cart is very simple, so that minimum maintenance is required

Bearing have been provide to reduce pulling force

9) Breaks have been provided for smooth movement of carts

10) Breaks are flexible enough to operate from front or rear and in combination or in isolation

11) Metallic wheels can be interchanged to pneumatic wheels with minor modifications

12) Yoke is of bamboo wood to reduce injury on animals shoulder caused by overheating

13) The cart is aesthetic and may be coloured to operator's choice

b) Briefly explain what is an Embodiment Design?

Ans. The Embodiment Design

A principal solution is already a first design proposal, because it embodies decisions on the geometry and material of the new product. It is, however, not more than an outline design proposal, which deals with physical feasibility only. It is a technical possibility that has to be worked out to some extent, before it can be evaluated against non-technical criteria as well. The development of a principal solution to a embodied design (see figure 3) can be seen as a process of establishing increasingly accurate, and more numerous characteristics of the new product, in particular: (1) the structure of the entire product (the arrangement of the parts) and (2) the shape; (3) the dimensions; (4) the material (s); (5) the surface quality and texture; (6) the tolerances and (7) the manufacturing method of all the parts.

A product design is ready for production once all the design properties have been specified definitively and in all required detail. Usually many properties have to be considered, and the relations among them are complex. Therefore the development of a principal solution into a detailed definitive design usually requires some stages in between. Typical intermediate stages are the design concept and the preliminary design (or sketch design).

In a design concept a solution principle has been worked out to the extent that important properties of the product - such as appearance, operation and use, manufacturability and costs – can be assessed, besides the technical-physical functioning. One should also have a broad idea of the shape and the kinds of materials of the product and its parts.

A preliminary design is the following stage and also the last stage before the definitive design. It is characteristic of this stage that the layout and shape and main dimensions have been established for at least the key parts and components of the product, and the materials and manufacturing techniques have been determined.

The modes of existence of a design proposal as described above, enable designers to explicate their thoughts about a design, and to judge and further develop them. Often there corresponds a more or less usual form of representation to each stage, such as flow diagrams for function structures, diagrams for solution principles, sketches for concepts, layout drawings for preliminary designs and standardized technical drawings for definitive designs. Such documents mark a stage in the development of the design and a phase in the design process.

In this phase the chosen concept is elaborated into a definitive design, also called definitive layout. The definitive design defines the arrangement ('layout') of assemblies, components and parts, as well as their geometrical shape, dimensions and materials ('form designs').

Contrary to what the phrase 'definitive' may suggest, the definitive design need not be completely worked out into full detail. The configuration of the product and the form of the parts are to be developed up to the point where the design of the product can be tested against all major requirements of the specification, preferably as a working model or prototype.

The decisions to be taken about the layout and form of the components and parts are strongly interrelated. Therefore, more than conceptual design, embodiment design involves corrective cycles in which analysis, synthesis, simulation and evaluation constantly alternate and complements each other. Embodiment design is essentially a process of continuously refining a concept, jumping from one sub-problem to another, anticipating decisions still to be taken and correcting earlier decisions in the light of the current state of the design proposal. It proves therefore difficult to draw up a detailed plan of action for this phase, that holds in general.

In Pahl and Beitz' model embodiment design is subdivided into two stages. The first stage is leading to a preliminary design, in which the layout, form and material of the principal function carriers are provisionally determined. In this stage several alternative embodiments of a concept are often worked up in parallel in order to find the layout. In the second stage, then, the best preliminary design is elaborated, up to the point where all major decisions about the layout and form of the product are taken and tests of its functionality, operation and use, appearance, consumer preference, reliability, manufacturability and cost can be carried out. Normally at the end of this phase the design is represented by layout drawings, made to scale and showing important dimensions, and preliminary parts lists.

The module structure specifies the division of a principal solution into realizable parts, components or assemblies, which has to be undertaken before starting the process of defining these modules in more concrete terms. Such a breakdown is particularly important for complex products, as it facilitates the distribution of design effort in the phase of embodiment design.

6. a) What is product design cycle? In your opinion which areas should be given importance in Engineering design

   i) Materials

   ii) Technology and

   iii) Marketing.

   b) In what ways the customer feedback is important in improving the design of an existing product or in launching a new product? Explain with the help an example. www.rgpvonline.com

Answer

Example of customer feedback to a website.

As a professional web design business, we have created and re-created many sites. The "great" ones have certain things in common. We thought we'd share those elements with you...

1. Easy to read. If background colors or images are used, the text on top of the background should be in a color that can easily be seen. Use a color scheme that complements and is pleasing to the eye. White space between images and sections of text make a page easier to view.

2. Easy to navigate. A visitor should be able to find the information they are looking for without hassle and frustration. The site's navigation buttons should be grouped together. If image links are used, text links should also be provided for those people who have images turned off on their browser or are using an older browser that doesn't support images.

3. Comfortably viewed. A Web site should be easily viewable in all screen sizes without a visitor having to scroll horizontally (left to right).

4. Quick to download. Graphics and sounds add download time to a Web page. Use them sparingly. Don't make your visitors wait too long for your site to download or they will click away and probably won't return. It is a good idea to find out what the approximate download times are for people who are using 28K and 56K telephone modems. Not everyone has DSL or cable Internet.

5. Avoid dead links. Make sure that links on all your pages are working, whether they are internal links to pages within your site, or links to external Web sites.

6. Keep the content fresh. People are more apt to return to your Web site if they find new and interesting material. Post articles on your site, offer a newly updated "Internet Special" or provide fresh, helpful links. All these things cause visitors to bookmark your site as a reference tool.

7. Clear and to the point. Visitors should have a clear understanding of what your Web site is about when they visit. Studies have shown that people do not like to read computer screens, so keep your Web site copy interesting to read and to the point.

8. Keep your target audience in mind. Think about the people who would be interested in visiting your Web site. If you are designing a web site about razor blades and shaving cream for men, the site should have a masculine feel to it. Decorating the page with pink hearts and roses would not be a good idea!

9. Provide a form for visitors to contact you. Visitors are more likely to fill out a form to contact you than clicking on an e-mail link. Always make things easy for your visitors... especially contacting you.

10. Browser compatible. Check your Web site in the most popular browsers to make sure everything is displayed properly. The top two browsers used are Internet Explorer and Netscape Navigator, but there are others such as the AOL browser, Mosaic, Opera and Web TV to name a few. Various versions of the same browser also display differently. It is a good idea to have a program on your computer that checks browser and version compatibility.

Here are some tips:

- Use correct labels for all fields
- Follow good form design principles
- Try to keep the number of fields to a minimum
- Offer tooltips and suggestions
- Display on-screen message on completion
- Use correct validation

CONCLUSION

Website usability plays a vital role in the success of a website. Good usability helps to provide a seamless experience for visitors and improves your chances of success. It is one of the factors that sets a professionally designed website apart from the rest. Given above are 11 usability characteristics that every website must display. It can help to make your website a success.

1. a) Explain what do you mean by user- centric design?

2. b) Explain with help of a free hand sketch any IT based design or any Graphic design that can reduce the daunting and repetitive task of grocery shopping easy.

Sample Answer:

1. More and more consumers are willing to share personal preference retailers must take advantage and engage consumers in conversations – through social media and direct feedback. This would help retailers better understand shopper expectations and preferences.

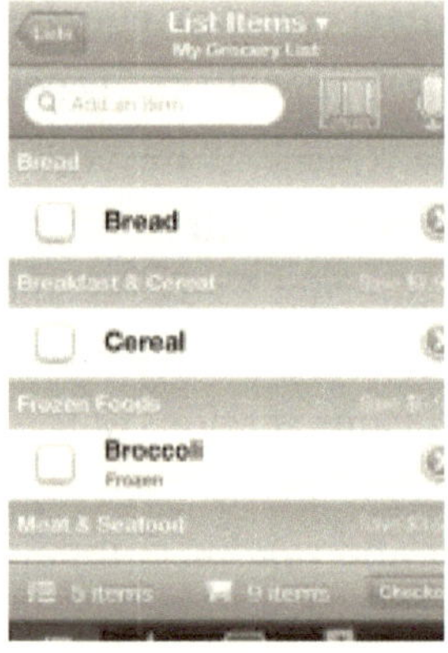

Fig. Grocery iQ lets you share your list with other shoppers.

In addition, retailers must start incorporating relevant information from connected, wearable devices like mobile or smart watches. Personalization solutions must become contextual and must focus on overcoming some of the limitations with the current breed of solutions in the market:

2. channel specific, generic segmentation, "black box-based" approach (without any knowledge of its internal workings) and limited

business context. These steps will further enhance the individual and household profile and help retailers better predict shopper behavior.

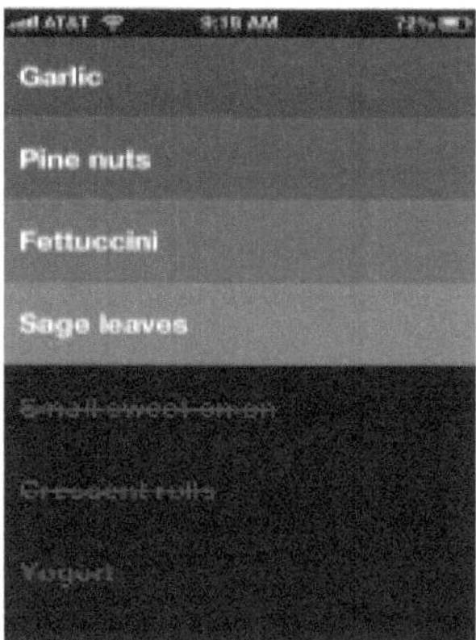

Fig. Clear offers an easy, no-frills way to make your grocery list

3. More often than not we see consumers start their shopping journey online, move into stores and complete their shopping in stores. Retailers must, therefore, not limit personalization to online but extend a seamless personalized experience to their physical stores as well.

4. One way to do this is to take advantage of the fact that shoppers are willing to identify themselves in stores. Retailers must then be able to facilitate personalization through in-store or personal shopper devices such as smart phones.Shoppers would have a delightful shopping experience as the device would guide them through aisles based on their shopping list orpreferences,

   – suggest new or complimenting products, provide

   – customized offers and also help them check out without waiting at the billing desk.

5. Customers must be provided with an option to have a part or the whole of their shopping delivered directly to their homes.

6. Retailers must also leverage store assistants with devices where important shopper preferences are made available to them in real time to deliver great service

7. Customers Clued In; they must not do personalization that customers should always feel that they are in control.

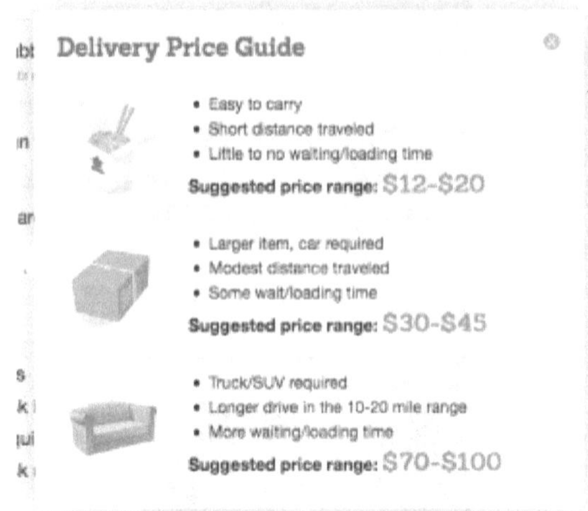

Fig. Too busy to even think about shopping? Hire a TaskRabbit to do it for you.

The onus is, therefore, entirely on the retailer to instill this confidence in customers.

Challenging economic conditions notwithstanding, the grocery sector is not immune to the "experience economy."

TheGroceryGame.com helps you find the best grocery deals in your area.

On the contrary, providing a personalized shopping experience, that is both seamless and memorable, is sure to resonate with the ever discerning grocery shopper. This will enable grocery retailers to sustain and expand their loyal customer base in the hyper competitive grocery market

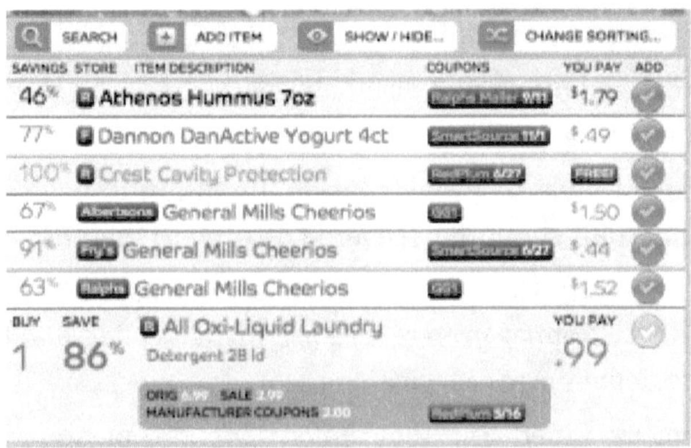